Techniques in Animal Nutrition Research

Techniques in Animal Nutrition Research

By :

A. Sahoo
S.K. Sankhyan
S.A. Karim

SSPH

SATISH SERIAL PUBLISHING HOUSE

403, Express Tower, Commercial Complex, Azadpur, Delhi-110033 (India)
Phone : 011-27672852, Fax : 91-11-27672046
E-mail :info@satishserial.com, hkjain1975@yahoo.com
Website : www.satishserial.com

Published by :

SATISH SERIAL PUBLISHING HOUSE

403, Express Tower, Commercial Complex, Azadpur, Delhi-110033 (INDIA)
Phone : 011-27672852 Fax : 91-11-27672046
E-mail : info@satishserial.com, hkjain1975@yahoo.com

Reprinted : 2015

ISBN 978-93-81226-05-6

Typeset at: Harminder *for* Laxmi Art Creation
Printed at: Salasar Imaging Systems, Delhi-110035

Preface

A wide range of methods and techniques are used in animal nutrition research and these methods should be well tested ones with adequate precision and reproducibility. Compilation of analytical techniques used in the field of animal nutrition research is required to provide comprehensive information to students and researchers in one place. This compilation "Techniques in Animal Nutrition Research" will be very useful as a reference book for the persons engaged in this field in their day to day work.

The research techniques were described under seven different heads, i) Common laboratory techniques ii) *In-vivo* feed evaluation iii) *In-vitro* feed evaluation iv) Techniques in grazing experiments v) Analysis of chemical constituents in feed and plant materials vi) Analysis of antinutrients and toxic principles vii) Rumen biochemistry and enzymes assay viii) Blood biochemical assay ix) Analysis of milk constituents and x) Statistical methods. Alternate techniques have also been given for parameters that can be estimated with equal ease and precision by more than one method. Since validity of data obtained in any biological experiment depends to a large extent on normal and average values, some standard tables are given for reference. Useful data on preparation of buffers and solutions are given in the Appendix.

In this book, efforts have been made to cover a wide range of topics like *in-vivo, in-vitro* studies, unique aspects of grazing studies *viz*. forage sampling method diet sampling technique, digestion trial conduction under field condition and some rumen microbiological aspects right from rumen content collection and anaerobic culture techniques with rumen liquor of digesta, methods of isolation of rumen microbes, blood biochemical profile and estimation procedures *etc*. The techniques are documented in a holistic manner to meet the requirements/needs of students pursuing Post graduate and Doctoral studies.

The present attempt to bring out 'Research Techniques in Animal Nutrition Research' involving methodological and analytical experiences of nutritionists will ease every biochemists/nutritionists to refer the protocol and come out handy with valid results. My thanks are due to these colleagues of mine for their active cooperation in bringing out this manual. I have great pleasure in acknowledging the invaluable help of my young colleagues who have spent much of their valuable time and energy in assembling, editing and bringing out this manual in the present form. My grateful thanks are due to Indian Council of Agricultural Research for financial support in bringing out this manual.

We will consider our effort rewarding, if this manual is widely used by researchers, teachers, students and technicians and find it useful.

We would like to admit that the present manuscript may not be free from errors as committing mistakes is human. Therefore, we would be grateful, if, such mistakes are pointed out by the readers for further improvement.

Authors

Contents

Chapter 1

Common Laboratory Techniques

1.1. Laboratory Safety

Everyone working in the laboratory should be cognizant of the potential hazards they face while working there. Fires with organic solvents, acid and base bums, and toxic fumes and vapors are common hazards in almost any nutrition laboratory. Generally, lab safety is a matter of common sense, but there are several rules that must be followed. Each student is required to read this chapter, which contains a brief synopsis of safety procedures. In addition, students must read the Standard Laboratory Procedures handout that is attached to the end of this chapter.

The following material is provided as a brief summary and guide to lab safety. It does not replace assigned reading material, but gives an overview of some important points.

Laboratory neatness	• Clean and neat work areas avoid risk of damage to clothing and books and injury from spilled chemicals. Neatness also reduces fire hazard.
Laboratory conduct	• Fooling around in the laboratory can be hazardous.

- Keep the lab in its proper place and fun and games in their place.

Working with glassware

- Remove frozen glass stoppers with proper equipment.
- Broken or chipped glassware should be discarded.
- Properly support glassware with ring-stands and clamps when heating and use cork rings with round-bottom flasks.

Working with glass tubing

- Do not touch heated glass until it has time to cool. Hot glass looks just like cool glass.
- To remove stoppers from glass tubing or thermometers, grasp tubing close to stopper and push gently with twisting. Use water or glycerin for lubrication.

Laboratory dress

- Pull hair back and wear eye protection when required.
- Sleeves that are too tight prevent freedom of movement, whereas sleeves that are too loose may cause you to overturn apparatus or glassware.
- Aprons protect clothing from corrosive or staining chemicals.
- Gloves protect hands from corrosive chemicals. Handle hot objects with insulated gloves.
- Do not wear open-toe shoes that allow spilled chemicals or broken glass to come in contact with your feet.

Working with test tubes

- Gently heat solids or liquids in a test tube near the liquid or solid surface.
- Be prepared to remove the tube from heat quickly to prevent eruption.
- Never point a test tube or reaction vessel at another person.
- For safety and neatness, place test tubes in a rack.

Chemicals in the eye

- Rapid treatment is vital.
- Run large volumes of water over eyeball until medical help is available.

	• Wash with large volumes of water for at least 15 minutes. Alkaline materials in the eye are extremely hazardous. • Know the location of the emergency eyewash station.
Safety shower	• Use this for chemical spills or a fire victim. • Operate by pulling down on ring and keep the area near the shower clear at all times. • Remove clothing from area affected by spills.
Fire on clothing	• Do not run or fan flames. • Smother fire by wrapping victim in fire blanket or lab coat and use the shower or a carbon dioxide fire extinguisher.
Extinguishing a fire	• Using a fire extinguisher: Know its location, Remove from mounting, Pull pin, Squeeze lever, Discharge at base of flame, Report use and recharge • Use dry sand to extinguish burning metals
Unauthorized experiments	• Always work under instructor's or lab technician's supervision in the laboratory.
Eye protection	• Normal eyeglasses are usually not adequate. • Do not wear contact lenses in the lab. • Eye protection is especially important when working with corrosive materials and vacuum and high pressure apparatus.
Acid/alkali spills	• For acid spills, use solid sodium bicarbonate followed by water. • For alkali spills, wash with water followed by dilute acetic acid.
Handling flammable liquids	• Flammable liquids should always be stored in an approved storage cabinet. • Extinguish all flames in the area where flammable solvents are used, as vapors may travel to ignition source and flash back.
Types of fire extinguishers	• Rating: For ordinary combustibles; wood, paper, and cloth. For flammable liquids; oil, grease, and gasoline. For use on live electrical equipment.

	• Number on extinguisher (e.g., IOA:5B) denotes square footage the unit is capable of handling.
Handling mercury	• Mercury spills are very hazardous. Droplets should be picked up by suction and a mercury spill kit used to complete cleanup. • Notify lab technician immediately when mercury spills occur.
Protection from toxic gases	• Emergency air masks should be used. However, because our lab is not equipped with such masks, clear the area where gases are, and notify the lab technician.
Waste disposal	• Hot glassware or reactive chemicals should be discarded in a non-metallic container separate from paper and other flammable waste. • Test-tube quantities of hazardous liquids can be flushed down the sink with plenty of water. Contact lab technician for disposal of large quantities of hazardous materials or anytime you are not sure of how something should be disposed of.
Labelling chemicals	• All chemicals should be clearly labeled. • Do not use materials from unlabeled containers. Avoid contamination. Never return reagents to their container. Clearly label chemicals as you work.
Carrying chemicals and equipment	• Carry long apparatus such as tubing or burets, in an upright position close to the body. • Grasp bottles firmly with both hands and hold them close to the body. • Do not carry bottles by the neck. Use a bottle carrier when transporting chemicals any distance.
Transferring liquids	• Remember, Acid to Water. Do not pipette by mouth, use a bulb. • Use gloves when pouring corrosive liquids. • Use a funnel when filling a bottle or flask and prevent an air block by raising the funnel. • Pour hazardous liquids over a sink.
Fume hood	• Use a fume hood equipped with a safety glass when working with toxic or flammable materials.

Gas cylinders	• Protect cylinder valve with cap. Fasten cylinders securely. Transport cylinders on a hand truck, don't roll. • Do not drop cylinders. • Mark cylinders when empty.
Handling sodium and potassium	• Fire or explosion may result when metallic Na or K are exposed to water. • Store them under light oil. • Metal can be cut safely with a spatula on a paper towel. • Destroy residues with alcohol. • Cool if necessary.

1.2. Chemicals and Reagents : Preparation and Precuations

Common terms and their definition

Solution: A solution is a homogenous mixture of two or more pure substances whose composition may be altered within certain limits. The substances in a solution are known as its components. Solutions of two, three and four components are referred as binary, ternary and quarternary solutions.

Solvent and Solute: Solvent is the component of solution whose physical state is same as that of resulting solution, while the other component is called as solute. When both the components have same physical state as that of solution, then the component in excess is called as solvent and the other as solute. The amount of solute dissolved in unit volume of solvent is termed as concentration. Solutions containing relatively higher concentration of solute are called as concentrated solutions, while those containing relatively lower concentration of solute are called as dilute solutions. The strength of solution denotes the number of grams of a substance dissolved in 1000 ml of a solution. The strength in general, of a solution, is expressed as either g/L or in terms of normality (N). A standard solution is, therefore, a solution of known strength.

The concentration of a solution may be expressed in any of the following ways:

Mass percentage: It may be defined as the number of parts by mass of solute per hundred parts by mass of solution. For example, a 5% solution means that 100g of solution contains 5g of solute

Volume percentage: Volume percentage may be defined as the number of parts by volume of solute per 100 parts by volume of solution.

Normality (N): Normality of a solution is defined as the g equivalent of the solute dissolved per L of the solution. A solution having normality equal to one is called normal solution. Such a solution contains one g equivalent of solute per L of solution.

N = Mass of solute (g)/(g equivqlent mass of solute × L of solution)

A semi-, deci- and centi- normal solution contains 1/2 (0.50), 1/10 (0.10) and 1/100 (0.01) gram equivalents of solute per litre of solution, respectively. A IN solution of an acid contains 1.00797g of replaceable hydrogen ions per litre of solution while that of a base contains 17.00737g of hydroxyl ions per litre of solution. A IN solution of a precipitating agent contains a weight of precipitating ion equivalent to 1.008g hydrogen. The amount for IN solution of oxidising and reducing agents is calculated by dividing gram molecular weight with total valency change in the ion concerned.

Equivalent weight: An equivalent weight of a substance is that weight equivalent in reacting power to an atom of hydrogen. The equivalent weight of an acid is obtained by dividing molecular weight with basicity, where basicity of an acid is equivalent to number of replaceable hydrogen ions in one molecule of the acid. Whereas, the equivalent weight of base is obtained by dividing molecular weight with acidity, where acidity of a base is equal to number of replaceable hydroxyl (OH) groups in one molecule of the alkali. The g equivalent weight is the equivalent weight expressed in g or is that weight equivalent in reacting power to 1 g atom (1.00797g) of hydrogen. Milli-equivalent and gram-milliequivalent weights are 1/1000th of equivalent and gram-equivalent weights, respectively. The gram-equivalent weight of an oxidising substance ($KMnO_4$) is weight in g of that substance equivalent to 8g of available oxygen. The g equivalent weight of $KMnO_4$ can be obtained as shown below:

$$2KMnO_4 + 3H_2SO_4 \rightarrow K_2SO_4 \pm 2MnSO_4 + 3H_2O + 5O$$

Therefore, 316.0752g [i.e. 2 (39.1020 + 54.9380 + 4 × 15. 9994) of $KMnO_4$ liberates 5 atoms or 79.997g oxygen during oxidation in acidic medium. Say 80g oxygen is liberated from 316g KMnO4 or 8g oxygen from 31.6g, which represents the gram equivalent weight of $KMnO_4$; Or $2KMnO_4 = 5(O)$ by which gram-equivalent weight can be calculated as 158g (molecular weight) divided by 5.

Molarity (M): Molarity of a solution is defined as the number of g moles of the solute dissolved per L of the solution. A solution having molarity one is called molar solution that contains one mole of solute per L of solution. It is expressed as moles per L or moles per dL. Both molarity as well as normality change with change in temperature.

M = Mass of solute (g)/(g molecular mass of solute × L of solution)

$$\text{Normality} = \text{Molarity} \times \frac{\text{Molecular mass (solute)}}{\text{Equivalent mass (solute)}}$$

For acids, Normality = Molarity × Basicity of acid (number of H^+ ions furnished by each molecule of acid in aqueous solution)

For bases, Normality = Molarity × Acidity of a base (number of OH^- ions furnished by each molecule of base in aqueous solution)

Molality (m): Molality of a solution may be defined as the number of g moles of the solute dissolved in 1000 g (1kg) of the solvent. A solution containing one mole of solute per 1000 g of solvent has molality equal to one and is called molal solution and expressed as moles per kilogram (rnol/kg). Molality of a solution does not change with temperature.

m = Mass of solute (g)/(g molecular mass of solute × Mass of solvent)

Formality (F): Formality of a solution may be defined as the number of g formula masses of the ionic solute dissolved per L of the solution. Formality is used to express the concentration of the ionic solids which do not exist as molecules but exist as network of ions. A solution containing one g formula mass of solute per L of the solution has formality equal to one and is called formal solution. It changes with change in temperature.

F = Mass of ionic solute (g)/(g formula mass of solute × L of solution)

Parts per milliion (ppm): Solute in very minute amounts is expressed in parts per million. It may be defined as the number of parts by mass of solute per million parts by mass of the solution.

Titration : The process of bringing a measured volume of a solution of known concentration into reaction with the desired constituents (or its equivalent of another substance) is called titration. In other words adding of a solution of known strength to another in order to complete the reaction is known as titration. Since volume of a solution of unknown strength is measured, it is also known as volumetric analysis.

Mathematical relations in titrimetry

Volume × Normality = Gram equivalent weight (e.g. 1/2L 0.2N reagent contains, 0.5 × 0.2 = 0.1 g equivalent weight of solute)

Volume of known × Normality of known = Volume of unknown × Normality of unknown

Titre: The titre is the weight of solute contained in a ml of solution or the weight of any substance which will react with or equivalent to one ml of solution.

Standard solution: A solution of accurately known concentration is called as standard solution, which may be prepared directly (primary standard) or by standardization (secondary standard) through reaction with. a primary standard. The end point of a titration occurs when chemically equivalent amounts of reactants are brought together, as indicated by an abrupt change in colour as shown by an indicator, at the stoichiometric point of the reaction.

Stoichiometric point: The stoichiometric point is an equivalence point at which an equivalent of the reacting substance has been added, irrespective of the type of reaction involved.

End point: The end point in an acid-alkali titration is that point at which the titration is stopped, being shown by the colour change of the particular indicator used.

Indicator: Indicator is a substance that indicates the physico-chemical status of a reaction. They may be internal, external or self indicators. They are mostly organic compounds of high molecular weight. When dissolved in water or any suitable solvent, they behave either as a weak acid or weak base. Basic indicator posseses a coloured cation and acidic indicator posseses coloured anion. The internal structural rearrangement is responsible for colour change. Important indicators used in titrimetric analysis are given below. The suitability or otherwise of an indicator in any given titration depends upon the pH value at which the indicator shows its specific colour change.

Indicator	pH range	Titration	Acidic Colour	Basic Colour
Methyl orange	3.1 - 4.4	Strong acid and strong base. Weak acid and weak base Strong acid and weak base	Orange - red	Orange - yellow
Methyl red	4.2 - 6.2	Strong acid & weak base	Red	Yellow
Phenolphthalein	8.2 -10.0	Strong acid & strong base Weak acid & strong base	Colourless	Pink
Bromocresol green	3.6 - 5.2	Weak acid and weak base	Yellow	Blue

Acidimetry: Acidimetry is a reaction in which the amount of base in a sample is determined by titration with a standard acid.

Alkalimetry: Alkalimetry is the measurement of acid in a sample by titration with a standard alkali.

Preparation of solutions

Normal solutions

Primary standards: Primary standards are used for the standardization of solutions of unknown strength, e.g. N/10 Na_2CO_3 N/10 Oxalic acid. Since, sodium carbonate is highly hygroscopic, dry it in hot air oven at 100 ± 2°C for 3-4 h to make it anhydrous, cool in desiccators and weigh exactly for preparing the desired strength.

Calculation: Mol. wt. of Na_2CO_3 is 105.989 and its Eq. wt. will be 105.989/2 = 52.9945

So, for N/10 Na_2CO_3 weigh 5.29945 g, dissolve in few amounts of distilled water (H_2O) and make up the volume in a volumetric flask with H_2O up to the mark of 1L.

Secondary standards: Secondary standards are prepared by normalizing with the primary standards. e.g. N/7 H_2SO_4, N/7 NaOH.

Calculation

Molecular weight of H_2SO_4 = 98.0776; i.e. (2 × 1.0080 + 32.064 + 4 ×15.9994)

g equivalent weight = 98.0776/2 = 49.0388g

Based on Sp.gr. (1.84 for conc. H_2SO_4), 49.0388g of H_2SO_4= 49.0388/1.84 = 26.65 ml.

Taking purity as 98%, 26.65ml of pure H_2SO_4 will be present in 27.19 ml.

To prepare N/7 H_2SO_4 1L solution will have 3.885 ml. Add slowly 4 ml of H_2SO_4 from the side of a beaker containing half the quantity of required distilled water, with the help of measuring cylinder and make up the volume to 1000ml by transferring the beaker contents to a volumetric flask after cooling and by repeated washings to the beaker. It is standardized against 10ml of N/7 Na_2CO_3 in a conical flask in the presence of 1 -2drops of methyl orange with the help of a burette. The change of colour from orange-yellow to orange-red will indicate the quantity of H_2SO_4 solution of unknown strength consumed to neutralize 10 ml of N/7 Na_2CO_3. Calculate the strength of unknown H_2SO_4 solution using the equation N/7 H_2SO_4 to add either distilled water or Conc. H_2SO_4 depending upon the increase or decrease in the strength of the solution, respectively. Titrations are repeated until at least 3 consecutive readings are obtained. The standardized N/7 H_2SO_4 is then stored in a clean reagent bottle after labeling. Based on this procedure, sulphuric acid of different normalities can be prepared as per the requirement.

Percentage solutions

These solutions are expressed as g/v or v/v in 100 ml, e.g. 25% H_2SO_4 (v/v), 3% boric acid (g/v), 40% NaOH (g/v), 10% formalin (v/v). To prepare 25% H_2SO_4, take 255 ml of 98% H_2SO_4 and add slowly on the side of a beaker containing 500 ml H_2O. There will be heat production, so cool it and then make up the volume to 1L with H_2O. Similarly, for the preparation of 3% boric acid weigh 30 g boric acid and dissolve in small volume of H_2O (say 400 ml) and make up the volume to 1L with H_2O.

Ratio solutions

Ratio solutions are prepared from liquid chemicals and expressed as v/v, e.g. 1:4 liquor ammonia (NH_3OH), 1:2 HCl, 1:4 H_2SO_4. To prepare 1:4 NH_3OH, take 200 ml of NH_3OH (it is available as 25-30%

solution and is generally prepared by dissolving ammonia (NH_3) gas in H_2O) and mix with 800 ml H_2O. Similarly, 1:2 HCl can be prepared by taking one volume of HCl to two volumes of H_2O slowly, cool, label and store in bottle (e.g. 333 ml HCl with 667 ml H_2O).

Storing and preservation of standard solutions

1. The bottle must be kept tightly stoppered to prevent evaporation loss of solvent to avoid increase in solute concentration.
2. The bottle should be shaken before the withdrawl of a portion of solution to ensure uniform composition of both withdrawn portion and the remainder left in the bottle.
3. Portions of solution once withdrawn should be discarded and never be returned to the bottle to minimise the danger of contamination.
4. Some standard solutions must be protected from atmospheric gases such as sodium hydroxide solution which is effectively diluted by atmospheric C02 by carbonic acid formation.
5. Solution containg potassium permanganate and silver nitrate should be stored in dark glass bottles (amber coloured) or kept in dark when not in use to prevent light catalysed decomposition. Solutions containing sodium thiosulphate should be protected from bacteria induced decomposition.

Cleaning of glassware

1. Ordinary glassware should be thoroughly cleaned with washing soda or any detergent followed by washing with ordinary tap water and rinsing with distilled water. Care should be taken to remove previous markings on the glassware, if any while cleaning.
2. Cleaning of oil flasks, used while estimating ether extrat, should be clone by slight boiling with dilute alkali (NaOH) followed by same procedure adopted for cleaning ordinary glassware. Care, however, should be taken not to use any brush for cleaning inside the flask to avoid scratch formations.
3. Graduated glassware may be cleaned by initially keeping in chrom ic acid solution (Dissolve about 60g potassium dichromate in 300 ml tap water by thorough stirring and boiling, to which around 450ml of commercial sulphuric acid is added slowly after

cooling) kept in cylindrical jar for reasonable time followed by washing and cleaning as per ordinary glassware. Discard chromic solution when it develops green colour.

Drying of glassware

1. Ordinary glassware can be dried by keeping in hot air oven at low temperatures (70-80°C.
2. Graduated glassware (pipettes, burettes, measuring cylinders, volumetric flasks etc.) sholud never be dried in hot air oven as high temperature would change the volume for which they are graduated for.

Other general precautions

1. Laboratory floor, working tables and water sinks should be kept neat and clean and it should be well ventillated and provided with an exhaust fan to remove unwanted gases, fumes and smoke.
2. One should work fully protected in laboratory by wearing white drill aprons and shoes.
3. Store chemicals and glassware in alphabetical order in well protected cupboards/alimirahs.
4. Reagents should be properly labelled with date of preparation before placing them on shelf.
5. Systematic breakage record should be maintained.
6. Always use acid and alkali gloves while handling strong acids and alkalies.
7. While working in Kjeldahl digestion room, use fume protecting face mask to avoid inhalation of highly irritating sulphur dioxide fumes.
8. Distilled water bottles should be kept tightly corked to avoid absorption of atmospheric gases.,
9. Always add acid to water slowly from the sides of the container near the sink.
10. While opening liquor ammonia bottles, especially during summer season, cool it for some time in a freezer to avoid sudden spurt of ammonia gas accumulated in the bottle.

11. Set the balance and check the oscillation of pans before using analytical balances. After use the pan and platform of the balance should be cleaned with a camel hair brush, if any spillage of chemical, sample etc. weighed was found.
12. Proper record of usage of special equipment should be made in log book meant for it showing date, time and condition of the equipment.
13. Fire extinguishers should be provided in each laboratory.
14. Always work in company during odd hours for the fear of fire and electrical accidents.
15. All used filter papers and other materials should be deposited in waste baskets.
16. Smoking in laboratory premises should be restricted to avoid catching accidental firing by highly inflammable chemicals.
17. All observations should be recorded at least in duplicate.
18. Never pipette strong acids and alkalies with mouth. Always use adopter or rubber bulb or bulb pipette.
19. Never blow the' solution left at the tip of the pipette and delivery of the reagent drawn into pipette should be uniform giving approprite time, varying from 10 to 30 sec. for quanties of 2 to 50 ml.
20. Mouth should be washed quickly with water or weak solution of washing soda during accidental sucking of acids.
21. Acid and alkali spillage on working tables, floor and clothes should be thoroughly washed with water after suitably neutralizing with either weak alkali in case of acid and weak acid in case of alkali.
22. Use always glass distilled water while analysing minerals.
23. Commercial sulphuric acid should only be used for digestion of samples for nitrogen/protein estimation.
24. No need to use distilled water while making up the volume of digested sample for protein estimation and during the analysis of crude fibre,
25. Always use self prepared reagents and indicators.

26. Consider lower meniscus for clear colourless and upper meniscus for coloured solutions while recording observations with the help of measuring glassware.

27. During cooling samples in a desiccator, the lid should be displaced to leave a small space, which can be closed after completely cooling.

28. Do not put on fans during decarbonization of a sample for ashing. Put on the exhaust during decarbonization and while handling fuming acids and other chemicals.

29. Always keep decarbonized samples in a closed container like desiccator while carrying to muffle, otherwise being light the material in silica basin may be displaced due to external air movement.

□□□

Chapter 2

In vivo Feed Evaluation

2.1. Measures of Digestibility

Nutritive value of feeds is determined by a number of factors, including composition, odor, texture, and taste. These factors are generally measurable in the case of the animal as digestibility and intake. Digestibility is simply a measure of the availability of food nutrients. When digestibility is combined with intake data, one can make a rather accurate prediction of overall nutritive value and, hence, the potential production a given feed can support. Of the two factors, intake is relatively more important than digestibility in determining overall nutritive value because highly digestible feeds are of little value unless consumed by the animal in question. However, digestibility usually provides a fairly reliable index of nutritive value because more digestible feeds are normally consumed to a greater extent than less digestible feeds. In addition, measures of digestibility are somewhat more easy to obtain than measures of intake, and, thus, considerable effort has been made by animal nutritionists to develop effective means of determining digestibility.

Several methods can be used to estimate digestibility, including the use of markers as described previously, harvesting forage and feeding it in a conventional digestion trial, collecting grazed forage via esophageal cannulae and feeding it in a conventional digestion trial, or collecting grazed forage via esophageal cannulae and determining its in vitro dry matter digestibility.

2.1.1. Conventional digestion trials

The conventional digestion trial is the most reliable method of measuring a feed's digestibility. Unfortunately, however, it is somewhat time consuming, tedious, and costly. Basically, the feed in question is fed in known quantities to an animal. Usually, the animal is restrained in an individual cage so that a quantitative collection of feces can be made. Accurate records of feed intake, refusals and fecal output are kept, and a subsample of each (usually 10% of daily output in the case of feces) is retained for analysis. When estimates of nitrogen balance are desired, urine output also is measured.

Three animals per feed are required as a minimum, with more animals preferred. The animals are usually allowed from 7 to 14 d to adjust to the feed and their cages, followed by a 5 to 7 -d period in which feces is quantitatively collected and feed intake is recorded. Samples can then be dried, ground, and analyzed for the nutrients of interest. Digestibility of any given nutrient can be calculated as follows:

Nutrient digestibility = [(nutrient intake - nutrient in feces)/ nutrient intake] × 100

Even though conventional digestion trials are the standard with which all other measures of digestibility are compared, the values obtained still vary ± 1 to 3 % as a result of animal-toanimal variation and sampling and analytical errors. Thus, it is clear why nutritionists would seek alternative measures of digestibility because of the time and expense of conventional trials. A typical protocol for nutrient balance trials is detailed in a short paper attached to this chapter.

2.1.2. Prediction of digestibility from chemical composition.

One such alternative measure of digestibility is the prediction of digestibility from chemical composition of the feed in question. This process involves development of multiple regression equations relating various chemical components to in vivo digestibility. Generally, the digestibility estimates obtained from prediction equations are not as precise as one might desire (±3 to 4% of values obtained from conventional trials), and at the present time, in vitro digestibility measurements are more extensively used to estimate digestibility than are prediction equations based on chemical composition.

2.1.3. Indirect measures of digestibility

The use of indigestible markers in nutrition studies

Digestibility and intake of feedstuffs have a considerable impact on the performance of livestock. However, conventional trials for determination of digestibility and intake are time consuming and expensive, and under certain conditions, may not provide valid estimates of these measurements. One important example of a situation in which conventionally determined values may be suspect is the grazing ruminant. A conventional approach to determining digestibility and voluntary intake of grazed forages would be to harvest the forage and feed it to penned animals. Unfortunately, this method ignores the animal's ability to select specific plants or portions of plants while grazing. Thus, harvested forage may not be representative of the forage actually consumed by the grazing animal. Because of such problems, indigestible markers or reference substances have been employed extensively in grazing research for determination of digestibility and intake. In addition, markers can be used to determine the rate of passage of nutrients through the gastrointestinal tract, site and extent of digestion, and microbial protein synthesis in ruminants.

The ideal marker: The requirements for an ideal marker can be briefly stated as follows:

- Should be inert, with no toxic physiological effects on the animal or micro flora.
- Should not be absorbed or metabolized within the gastrointestinal tract.
- Should be physically similar to or intimately associated with the material it is to mark.
- Should not influence gastrointestinal secretion, digestion, absorption, or motility.
- Should have physiochemical properties that allow for precise, quantitative analysis, and it must not interfere with other analyses.

Unfortunately, none of the markers currently in use satisfy all these criteria. However, with proper selection of a marker, based on specific experimental conditions, effective measurements can be made. For example, chromium sesquioxide can be used to measure digestibility, even though it does not remain intimately associated with either the particulate or fluid fractions of digesta because it is only necessary that it be fully recoverable (i.e., non-absorbable) for digestibility measures. It could not, however, be used as an index of digesta flow rate. A number of markers are presently used in nutrition studies. These can be divided into internal and external markers. Internal markers are integral components of the feedstuff, and lignin is the most widely used. Recently, there has been considerable interest in the use of acid insoluble ash as an internal marker. Indigestible ADF and NDF also have been studied and seem to have considerable promise.

External markers: These are indigestible substances added to a feedstuff. Chromium sesquioxide ($Cr_2 0_3$) is chief among external markers, but several other markers including cerium, dysprosium, ytterbium, and a ruthenium phenanthroline complex have received considerable attention. Many of the rare-earth element markers like Yb are made to bind with a specific feed, and as such, are quite useful in determining flow rates of particular feed ingredients. In ruminants, it is often desirable to determine the quantity of microbial protein synthesized with various diets because microbial protein makes a significant contribution to the nitrogen needs of ruminants. It is possible by using markers mentioned above, to determine the total protein passing to the lower tract in ruminants; however, it is then necessary to determine what portion of the total protein is of microbial origin. This can be estimated by using a number of markers specific to bacteria, including diaminopimelic acid (DAP), ribonucleic acid, and purines. Diaminopimelic acid is thought to be found only in bacteria, so by determining the DAP content of digesta and knowing the DAP:N ratio in bacteria, one can estimate the portion of nitrogen in the digesta of microbial origin. In the case of RNA, it is assumed that feed RNA is broken down in the rumen and only RNA of bacterial origin passes to the lower tract. Again, when the RNA:N ratio in bacteria is known, the microbial contribution to total protein in the digesta can be determined by measurement of digesta RNA concentrations. It is to keep in mind that DAP and RNA are not indigestible markers, but serve only to aid in determination of the microbial protein content in ruminant digesta.

Use of markers to estimate digestibility

Estimates of total tract or specific site digestibility are possible with many of the markers described above. The digestibility of a given nutrient can be determined by application of the following formula:

$$\text{Nutrient digestibility (\%)} = 100 - 100 \times \frac{[\text{\% marker in feed}]}{\text{\% marker in feces}} \times \frac{\text{\% nutrient in feces}}{\text{\% nutrient in feed}}$$

This equation can be used with both internal and external markers. For example, lignin could be used as an internal marker and percentage of lignin in feed and feces could be determined, as well as the percentage of the nutrient in question. This method allows digestibility to be calculated without total collection of feed and feces; that is, grab samples of feed and feces can be used for analyses. Chromic oxide is probably the most widely used external marker for this particular application, with a known percentage of Cr_2O_3 being incorporated into the feed. The detail procedure is given in another chapter.

$$\text{Ruminant nutrient digestibility (\%)} = \frac{[\text{\% marker in feed}]}{\text{\% marker in duodenum}} \times \frac{\text{\% nutrient in duodenum}}{\text{\% nutrient in feed}}$$

Use of markers to measure intake

As we have discussed before, intake has a significant effect on the economy of livestock production. Intake of the grazing ruminant is difficult to measure by conventional means, since harvested forage may not be representative of forage actually grazed. Thus, researchers have applied indicator methods to estimate intake. When fecal excretion and digestibility of the diet are known, intake can be readily calculated. The detail procedure is given in another chapter.

It is given in this calculation that the quantity of fecal excreta produced by an animal is a function of intake and indigestibility of a particular diet. This expression can be written in equation form as:

Fecal output (g) = DM intake × % DM indigestibility

Rearranging this equation, one can solve for dry matter intake as follows:

Dry matter intake (g) = Fecal output (g)/ % Dry matter indigestibility

Now the problem becomes one of estimating fecal output and digestibility. Fecal output is usually estimated with indigestible markers, but can be determined by fitting grazing animals with fecal collection bags and conducting a total collection of feces. Chromic oxide is probably the most frequently used marker to estimate fecal output, and it is normally administered as a bolus twice daily (e.g., 0800 and 1700 h) for a preliminary period of approximately 7 d, followed by a collection period of approximately 5 d. During the collection period, fecal "grab samples" are obtained via rectum at the same time as bolusing. Chromic oxide can then be determined by wet ashing fecal samples followed by atomic absorption or spectrophotometric measurement of chromium content. Fecal output is then be calculated from the following formula:

Fecal output (g) = (Indicator consumed (g/d)/(Indicator concentration in feces (g/g of DM)

A potentially significant problem in determination of forage intake with Cr_2O_3 or similar markers is the diurnal excretion pattern of such markers. That is, fecal excretion is inconstant, or variable throughout the day. As such, fecal excretion of a marker may not be constant during the day and over- or under-estimation of chromium concentration in fecal grab samples can occur. Nevertheless, a large amount of data on forage intake has been collected by this method and the technique seems quite useful, especially when relative, rather than quantitative, estimates of intake are desired.

Use of markers to measure rate of passage

Rate of passage of particulate and fluid materials through the digestive tract is of considerable importance in livestock production. In ruminants, fluid flow rate, or more specifically fluid dilution rate (percentage of fluid volume leaving the rumen per unit time) has been shown to be related to the efficiency of microbial growth. Generally, the faster the dilution rate, the more efficient is microbial growth. Rate of passage of particulate material from the rumen can have a significant effect on intake and digestibility. Grinding and pelleting of forages tends to increase rate of passage, allowing for an increased voluntary consumption of feed, as bulk fill no longer limits intake. At the same time, digestibility is often decreased by grinding and pelleting forages because particles have less opportunity to undergo fermentation. Moreover, changes in digestibility may effect rate of passage. If feed is fermented rapidly to small particles, rate of passage may increase, and subsequently, intake will increase.

Rate of passage of specific nutrients can also be calculated. This is an alternative to partitioning digestibility as described previously, in that the passage of a specific nutrient from the rumen or other organ could be used to "back calculate" digestibility when intake is known. Several markers are available to measure the rate of fluid passage from the rumen. Perhaps the most commonly used are chromium EDTA, cobalt EDTA, and polyethylene glycol (PEG). One method involves continuous infusion of the marker with time sequence sampling. For example, the marker could be given in the diet or directly infused into the rumen until an equilibrium point had been reached. Equilibrium can be defined as the point when marker concentration at any sampling point is constant. When equilibrium has been reached, samples can be taken at several intervals and combined. Numerous samples are taken in order to overcome the effects of individual "nonrepresentative" samples.

The flow rate of digesta past a given sampling point can be calculated as:

Flow rate at a sampling site = Infusion rate/Marker concentration at sampling point

For example, if 10 g of PEG are infused into a sheep and the marker concentration at the abomasum is 0.01 g/mL, then the flow rate past the abomasum (from the rumen) would be calculated as 10 g/d/0.01 g/mL = 1,000 mL/day. The flow rate of a given constituent of the fluid, could be calculated by multiplying the flow rate of the fluid by the concentration of the constituent in the fluid at the sampling site. This method is quite useful; however, it does not allow the volume of a specific pool to be calculated, simply the flow rate. A means by which both flow rates and volume can be calculated is the single-dose method with time sequence sampling. In this technique, a known amount of marker is dosed into a particular segment of the digestive tract, and its concentration in digesta is determined by sampling at time periods following the dose. If, for example, one wished to determine the fluid volume and outflow rate of the rumen, CrEDTA could be administered through a ruminal cannula and ruminal fluid samples taken at periods following the dose. Rumen volume is assumed to remain constant, with inflow equal to outflow. When the Cr concentration in the fluid is plotted against time, a non-linear curve will result. Plotting the natural log of the Cr concentration against time, however, will yield a linear relationship. The slope of this log curve equals the fractional dilution rate or, in other words, the percentage of volume passing per hour. Extrapolation to concentration at time zero and division of this value into the dose yields an estimate of the volume. It is to be noted that this method can be applied to estimate the volume and fractional dilution rate of any pool (e.g., plasma free fatty acids, body glucose, ruminal VF A, and so on) if an appropriate marker is chosen. An alternative calculation method to log transformation is the use of non-linear regression techniques.

Measurement of particulate passage rate: It can be accomplished by the same methods described above. In other words, a continuous or single dose of a particulate phase marker can be given and flow rates and/or volumes calculated as described above. Chromic oxide has been used for this purpose; however, it has the disadvantage of not being intimately associated with the particulate phase. More recently, ytterbium, dysprosium, cerium, and a ruthenium phenanthroline complex have been used to estimation of particulate turnover rates. These materials seem to bind tightly with particulate matter and closely meet the criteria for ideal markers described previously. In the case of rare-earth elements, feedstuffs are usually allowed to soak with the rare earth for some time, followed by washing off of unbound rare earth. Treated feedstuff can then be fed or dosed for estimation of passage rate. Many of these markers are used as the radioactive form, in an effort to facilitate measurement; however, animals treated with radioactive substances fall under strict federal guidelines. Two-compartment models are also proposed to describe the fecal excretion of particulate markers, which allows determination of particulate passage rates.

Internal markers like indigestible ADF can also be used to determine rate of passage. As yet, the author is not aware of studies in which such markers as acid insoluble ash have been applied to the estimation of rate of passage.

Markers for measurement of microbial protein synthesis

As discussed earlier in this chapter, it is often desirable to determine the amount of microbial protein synthesized by ruminants. This can be accomplished with the use of such markers as diaminopimelic acid, RNA and purines. How microbial protein synthesis can be estimated by the use of purines as a microbial marker is described elsewhere.

2.1.4. *In situ* (Nylon Bag) digestibility techniques

Another method of estimating digestibility of feeds is the nylon bag technique. In this procedure, nylon bags (~5 cm × 15 cm) are filled with 2 to 3 g of the feed in question and incubated in the rumen of a cannulated animal. Generally, bags are secured to a weighted cord to prevent floating in the rumen and to ensure adequate exposure to microbial digestion. Bags are then removed, washed under tap water (until wash water is clear), dried (50°C for approximately 24 h, followed by 1 h at 100°C), and the weight of residue determined. An empty bag should be incubated and serve a similar purpose to the blank tube in the Tilley and Terry procedure. A 72-h period in the rumen often provides estimates of digestibility similar to those obtained by the Tilley and Terry method (described in the later section).

Many of the same variables (i.e., diet of animal, time of fermentation) that affect the Tilley and Terry method also affect the nylon bag procedure. In addition, one should consider the pore size of the nylon material, which should be small enough to prevent passage of feed from the bag, but large enough to permit microbial entry. A pore size of 50 urn or less is desirable. Furthermore, the sample:bag size ratio is quite important, and a ratio of ~10 mg/cnr' of bag surface is probably adequate. One disadvantage of the nylon bag technique is that fewer samples can be run at one time than with the Tilley and Terry method, and a donor animal with a large diameter cannula is desirable. Nylon bag (or in situ) techniques, are, however, quite useful for evaluating kinetic aspects of digestion in ruminants. Through the use of multiple incubation times and computer models, rates of nutrient digestion can be estimated.

Procedure

1. The nylon bags are cleaned by turning them inside out and then washing in a mild detergent solution and tap water.
2. The bags are kept immersed in water for 3-4 h and washed again in running tap water and then dried at 60°C for 48 h.
3. About 4-5g of air dried sample is taken in each bag, tied separately with nylon string. The other end of string is tied to a nylon rope, with length of about 1.5 m. To one end of nylon rope is attached 200g weight and other end is secured by encircling it around the flank of the animal.
4. For each sample, say thirty six bags (12 in each animal) are suspended in rumen of fistulated animals. The bags are soaked in water for about 1 min. before incubation in rumen.
5. They are withdrawn after 1, 2, 3, 6, 9, 12, 24, 36, 48, 60, 72 and 96 h.

6. On removal, the bags are held in running tap water to remove adhering digesta from the outer surface of the bags, given a gentle squeeze and kept under running tap water till the out-coming fluid was clear.
7. The bags are dried to a constant weight in a forced draft air oven at 65°C for 48 h, weighed and DM loss estimated for each incubation hour.
8. The residual DM at each incubation hour is analysed for ash, N and NDF. Nitrogen is also estimated in the residual NDF. Zero time DM washing loss is estimated by soaking 3 bags per sample in warm tap water (39°C) for 1 h followed by washing and drying.

Several suggestions for a standardized procedure are summarized by Vanzant et al. (1998) and presented below for ready reference.

Item	Recommendation
Diet	
Type	60-70% roughage
Feeding level	Maintenance
Feeding frequency	≥2 times/day
Bag	
Material	Polyster
Pore size	40-60 μm
Sample size: surface area	10mg/cm^2
Sample processing	2 mm screen size (willey mill)
Replication	
No. of animals	≥2
No. of days	≥2
No. of bags	≥1
Incubation procedure	
Preincubation	Not necessary
Ruminal position	Ventral rumen
Insertion/removal	Removal simultaneously
Incubation times	To describe curve
Rinsing	
Microbial correction	Yes
Marthematical model	Simplest available to adequately describe data
Standard substrate	yes

Use of internal standards

Because of temporal variation of in situ measurements, the use of an internal standard is critical. The choice of feedstuffs to use for these standards is left up to the individual researcher. However, this selection should be made while considering that the standards should represent the feedstuffs of interest as closely as possible. Thus, a variety of standards representing a wide range in chemical composition and in situ disappearance values may be required.

Calculation and interpretation of results by mathematical models

To estimate the DM, OM, CP, NDF and NDIN degradation constants three models were used a) Log natural transformation and linear regression (LT-LR), b) Direct non-linear least squares method (Orskov & McDonald model) and c) Gompertz model.

Logarithmic transformation and linear regression : Logarithmic transformation is used by determining the natural logarithm of potentially digestible residue remaining at the end of each fermentation period and regressing upon time. Potential extent of digestion of DM, OM, CP, NDF and NDIN is the value for respective disappearance at the end of fermentation (96 h) which is taken as 100% for further calculations. The value for disappearance at each preceding hour is adjusted accordingly. Potentially digestible residue at different incubation hours is estimated by subtracting potential extent of digestion at that particular hour from 100. Degradation rates for all feeds tested are calculated assuming there was no undegradable fraction or a+b = 100.

Here, (a + b) represents total degradability which is asymptote of degradation curve. It follows that 100 - (a + b) is the fraction which is totally undegradable in the rumen. The equation is constrained so that (a + b) can not exceed 100. If 'a' is found to be negative, indicative of a lag phase, the value for lag is estimated (Orskov, 1982). The equation for estimation of lag time is as follows:

$$PD_a = PD_p \times e^{-c \times t}$$

Where,

PD_a = potential degradability observed at 0h

PD_p = potential degradability predicted by linear regression

c = rate constant

Solving for 't' determines the time (h) at which PDa = PDp assuming 'c' rate constant for digestion, 't' then is the lag time for digestion. Once lag time was determined, DM or CP or NDF residues from 0h through lag time (inclusive) are deleted from the data set and regression coefficient (-c) redetermined on the remaining data points (Smith *et al.*, 1972; Mertens, 1977).

Direct non- linear least squares (DNLS): Curve fitting and parameter values are estimated by the Marquardt's Compromise (Draper and Smith, 1981) of non-linear regression method using user defined non-linear regression of Sigma-Stat for Windows (Jandel Corporation, USA, 1995). Up to 200 iterations are used to fit an equation. When the R^2 value did not change in the fifth significant Figure for five consecutive iterations, convergence is assumed. Starting parameter values are taken from literature. The options used are that a) the algorithm minimised the sum of squares of the actual distance of the points from the curve, b) the derivatives are calculated with a slower, more accurate method and c) the convergence is reached when two consecutive iterations chanced the sum of squares by less than 0.01%. The degradation kinetics of the incubated feedstuff is described by curvi-linear regression of DM, OM, CP, NDF and NDIN loss from the bag with time. This model is also known as simple negative exponential curve (monomolecular, Mitscherlich or first order kinetic model) with lag phase (Orskov and McDonald, 1979).

A Mitscherlich equation (first order exponential model, Orskov and McDonald, 1979) is used to fit the disappearance data obtained from nylon bag incubations. The model is described by the following equation:

$$P = a + b^*\left(1 - e^{-c^*(t-L)}\right)$$

where,

P = potential degradability

t = incubation time

a = Y axis intercept at time 0. Represents soluble and completely degradable substrate that is rapidly washed out of the bag

b = the difference between the intercept and the asymptote. Represents the insoluble but potentially degradable substrate which is degradable by the microorganisms according to first order kinetics

c = rate constant of b function

L = discrete lag time

The DM, OM, CP NDF and NDIN degradability data for the various times of incubation for each sample are also fitted with the Gompertz model (Bidlack and Buxton, 1992), without considering the washing losses.

$$y = B \times e^{-C \times e^{-A \times t}}$$

where,

y = fraction degraded at time t
B = asymptotic value (total potentially degradable fraction)
C = relative degradation rate
A = constant factor of microbial efficiency

Fractional degradation rate (?) at $t_{½}$ is calculated as:

$$\mu = C + \frac{A}{2\sqrt{t}}$$

Immediately available fraction (IAF) is calculated by setting t equal to zero in the Gompertz model. According to this model, fractional degradation rate varies as a function of time and an average value (i.e. a constant value comparable to the exponential rate of degradation) can be derived as:

$$C = \frac{A}{C}$$

Once the constants 'a', 'b' and 'c' were known, the effective degradability is calculated by the equation:

$$ED = a + \frac{b*c}{c + k_p}$$

where k_p = rumen small particle outflow rate

The ED is calculated at different assumed kp of 0.02,0.05 and 0.08 per hour corresponding to the passage rate at maintenance, growth and lactation (AFRC, 1992). Coefficient of determination (R^2) was estimated to know about the goodness of fit of the data to the model (Draper and Smith, 1981). Digestible undegraded nitrogen (DUN) or truly absorbed DUN content is calculated as per Webster *et al.* (1986) by the equation:

DUN = 0.9 (UDN- ADIN)

Quickly (QDP), slowly degradable (SDP), effective rumen degradable (ERDP) and digestible undegraded (DUP) protein is also calculated as per the equations given by AFRC (1992).

$$QDP = a * CP$$

$$SDP = CP \times \frac{b \times c}{c + k_p}$$

$$ERDP = 0.8 * QDP + SDP$$

$$DUP = 0.9 (CP- QDP- SDP- 6.25 *ADIN)$$

Over the last two decades, in situ techniques have been used extensively for measuring ruminal degradation of feedstuffs. Current predictive models put renewed emphasis on the need for quantitative information regarding rates and extents of ruminal degradation. However, in situ techniques suffer from tremendous variation, both within and among laboratories. A considerable number of studies have evaluated the influence of various factors on in situ-derived estimates of ruminal degradation. Factors that should be addressed in a standardized procedure include:

i) bag and sample sizes

ii) bag material and pore frequency

iii) bag insertion and removal procedures

iv) location of bags within the rumen and containment

v) procedures for the bags

vi) rinsing procedures

vii) microbial correction

viii) incubation times

ix) mathematical models

x) numbers of replicate animals, days, and bags required to obtain repeatable estimates of ruminal degradation.

Currently, the lack of standardization in rinsing techniques and the failure or inability to correct for microbial contamination of in situ residues seem to be the major sources of variability with in situ procedures.

References

AFRC (1992) Institute of Arable Crops Research Annual Report 1992. Agriculture and Food Research Council, UK.

Bidlack, J.E. and Buxton, D.R. 1992. Content and deposition rates of cellulose, hemicellulose and lignin during regrowth of forage grasses and legumes. Can. J. Anim. Sci. 72: 809-818.

Draper, N.R. and Smith, H. 1981. Applied Regression Analysis. John Wiley and Sons Inc., New York.

Mertens, D.R. 1977. Importance and measurements of protein insolubility in ruminant diets. Proc. Georgia Nutr. Conf. pp. 30.

Orskov, E.R. 1982. Protein Nutrition in Ruminants. Academic Press, London, U.K.

Orskov, E.R. and McDonald, I. 1979. The estimation of protein degradability in the rumen from incubation measurements weighted according to rate of passage. J. Agric. Sci. (Camb.). 92: 499-503.

Smith, L.W., Goering, H.K. and Gordon, C.H. 1972. Relationship of forage compositon with rates of cell wall digestion and indigestibility of cell walls. J.Dairy Sci. 55:1140-1147.

Vanzant, E.S. Cochran R.C. and Titgemeyer E.C. 1998. Standardization of in situ techniques for ruminant feedstuff evaluation. J. Anim. Sci. 76:2717–2729

Webster, A.J.F., Kitcherside, M.A., Keirby, J.R. and Hall, P.A. 1986. Evaluation of protein foods for dairy cows. Anim. Prod. 38: 549 (Abstr.).

2.2. Balance Studies in Animal Nutrition : Assessment of Nutrient Loss in Metabolism Trials

The determination of nutrient balances is used extensively in nutrition research. It is obviously of great significance to know how much of a given nutrient in a feed is retained in body tissues. Theoretically, if all inputs and outputs are accurately measured, the difference between them will establish the amount retained. Thus, net balance should equal retention. In practice, net balance figures are usually higher than retention data and frequently represent unrealistic retention situations. It is therefore important that a tool so widely used and with such an impelling logical basis be carefully evaluated to discover its weaknesses, limitations and conditions influencing interpretation.

Significance

Knowledge of the quantitative needs of the body for different nutrients (carbohydrates, proteins, fats, water, minerals and vitamins) and the relative value of feeds as sources of them is the basis of scientific feeding. Balance studies which measure the amount of nitrogen or various minerals in the diet and the amount in the faeces and urine, have been used to estimate the adequacy of diets or to determine nutritional requirements. In the latter case, the nutrient intake is usually varied and an estimate derived of the amount required for nutrient balance. Theoretically, in the well nourished healthy adult animal who is not growing and who maintains approximately the same weight over long periods of time, body composition must be approximately constant and the intake of various nutrients must approximate the amounts that are lost from the body. However, in exceptional cases like trace minerals, which may accumulate or be lost over long periods of time since the amounts may not be significant quantitatively and might have no observable physiological effects. For nitrogen and major minerals, however, there are obvious limits and these can be evaluated relative to whatever is known about general body composition at different age and state of production. During pregnancy or growth, there must be positive nutrient retention and the nutrient requirements in these conditions, have often been estimated by calculating the approximate retention which would be required for the synthesis of new tissue expected over and above the amount of estimated maintenance requirement. In either case(s), pronounced retentions or losses must eventually produce abnormal physiological/pathological consequences.

Balance data for groups of nutrients such as ash or ether extract would be meaningless, although the balances of major minerals are often carried out in many nutrient balance studies. Similarly, balance studies of single organic species such as an amino acid, a carbohydrate, or a vitamin would also be meaningless since such compounds could be synthesized by intestinal microorganisms or degraded in the gut so the levels in the excreta do not represent feed nutrients not absorbed by the host. Therefore, balance studies of nutrients other than energy are essentially limited to nitrogen and the inorganic nutrients. Since energy balance is a specialized topic that deals with energy metabolism and accounted also the N balance data, the present paper concentrates on various issues related to N balance studies.

Excretory routes

The different excretory routes include principally, facal and urinary routes. In milk and wool producig animals and egg laying birds, the loss through these produce are also accounted in the balance studies. The additional routes of excretion are through the skin as sweat, desquamation of skin, hair, nails, etc and in saliva and expired air. Loss of nutrients in some of the pecuiliar situations like walloping, belching, regurgitation, vomiting, flatulence, etc. cannot be ruled out. These losses are not measured in the usual balance study, thus critically, the positive balances are overestimated. Attempts to measure such losses, however, indicate that under ordinary conditions they are of the order of 0.3 to 0.5 g N/day (Mitchell, H. H. and Hamilton, T. S. 1949; Galloway et al., 1971). If one accepts the higher value and subtracts this quantitated loss from the estimated balance value, the conclusions are not generally changed to any substantial degree.

2.2.1. Accounting N loss in metabolism trial

While assessing N loss in a metabolism trial, e.g. a 45-kg pig fed a corn-soybean meal diet may have an efficiency of N utilization of 35% (NRC, 1998). This low efficiency is the result of approximately 7% of the dietary CP not being digestible and the equivalent of 8% of the dietary CP being lost in endogenous excretions. The synthesis of endogenous material results in the obligatory catabolism of the equivalent of 10% of the dietary N. These losses account for 25% of the dietary N intake. The remaining CP (75% of dietary) is available for lean tissue accretion; however, as a result of mismatches between requirements and dietary supply, approximately 30% is degraded and used for energy production. The remainder is actually used for lean tissue growth, but approximately 10% is not accreted due to inefficiencies in lean tissue growth, and summation of these losses results in N utilization in the neighborhood of 35% (van Kempen and van Heugten, 2000). Feed wastage and poor animal health can deteriorate this efficiency even further. According to Chung and Baker (1992), efficiencies of 60% can be achievable in nursery pigs with diets formulated to be nearly 100% digestible. However, in many instances researchers using commercial diets report efficiencies of N accretion in the 60 to 70% range, which draws criticism on methodology accuracies accounting to various channels of N loss. According to NRC (1998), N accretion in pig, say 20 g, is equivalent to protein accretion of 125 g (6.25 multiple of N), which is equivalent to a lean tissue gain of 543 g/d (approximately 23% protein in lean body tissue on DM basis). In dairy cattle, the empty body weight (BW) gain is 95.6% of shrunk BW gain or average daily gain (ADG) and the efficiency with which net protein is used for gain is 0.28908 (NRC, 2001). Accordingly, the N content of gain is accounted as 30 g per kg gain in growing calves. This means, a N balance of 30 g in any experimental protocol involving growing calves should be equivalent to ADG of 1 kg, which most would agree is unrealistic. However, for simple calculation the protein content of animal body less the contents of digestive tract is 18-19% in many animal species except pig (=30 kg), sheep, steer (fat), which have lesser values (13-16%) and this also explains a similar conversion factor for ADG. Possible explanations for the very high N retention include incomplete collection of excreted N or unaccounted feed waste. Another explanation, however, is N loss from the urine. It is well known that N in the form of ammonia can volatilize quickly from urine contaminated with bacterial urease unless the urine is stored in a closed container, cooled, or acidified.

Urinary urea, following conversion to ammonia, is the N excretion product most vulnerable for loss to the environment and can have a profound effect on the interpretation of data in that ammonia losses from urine are influenced by many factors. For example, Shriver et al. (2003) used a test diet high in synthetic amino acids. This diet resulted in a drop in urine pH of 0.4 pH units. Ammonia emission is very sensitive to pH, as outlined by the authors. Thus, if ammonia losses were a problem in this trial, and if they were altered by urine pH, apparent N retention would be affected simply as a result of the change in urine pH. A regression equation is delineated for N utilization efficiency (NUE = 37.1 + 0.00048 x N intake [g/d, P = 0.95] – 0.058 * urine N excretion [g/d, $P < 0.001$]; s.e. obs. = 3.98; $r^2 = 0.59$) which indicates 37% mean theoretical maximum NUE for the animals used in the calculation, the remaining represents obligate losses through protein turnover and other maintenance requirements (Moorby and McConochie, 2002). The effect of N intake on NUE was found to be non-significant while the effect of the coefficient for urine N excretion was highly significant, which suggests that the major effects on NUE are those occurring within the animal which lead to losses of N in urine essentially regardless of the level of N intake. This emphasizes the importance of accurately assessing urinary N loss during a metabolism trial.

2.2.2. Factors affecting assessment of N balance

Nutrient balance studies have been used extensively in nutritional research. Several problems of determination have been observed, which include errors in collection and sampling of feed and excreta, loss vectors not measured, correspondence of excreta collections and feed intake. Such errors usually result in overestimations of actual balance and render absolute values questionable. Wallace (1959) stated that intake values always tend to be overestimated because the animal cannot eat more than is offered while excreta are always underestimated because it is not possible to collect more than 100% of the excreta. In addition, experimental design may also influence results. Balance data are influenced by levels of other nutrients, growth rate, depletion status of the animals, adaptation periods, and other factors (. Urinary N losses can have a profound effect on the interpretation of data in that ammonia losses from urine are influenced by many factors, which are outlined below.

Experimental design

It is obviously necessary that both balance and retention trials be conducted with skill and care, and that certain errors inherent in the conduct of balance trials will continue to be problems. However, the technical correctness of determination is only one factor that needs to be considered in evaluating the results of balance or retention experiments. If the balance trial is designed to reflect differences in the nutrient level or availability in feeds, the nutrient must be fed at less than its requirement, however, if balance studies are used to determine the comparative capacity of animals to utilize nutrients, the intake must be in excess of the requirement.

Duration of feeding trial

Forbes (1973) has suggested that the balance studies which measure the amount of N or various minerals in the diet and the amount in the faeces and urine may be erroneous because modification of body composition after a dietary change may occur slowly and exponentially with time. The usual short balance periods of 2 weeks or so would be too short to identify stable conditions, but the argument seems to be justified in Ca balance studies where high retentions have been reported at high intakes and the reason being adaptation to different Ca intakes may require very long periods. In conventional balance studies of limited durations, the usual finding is "The more that is fed, the more that is apparently retained". These retentions, even if they are real, cannot be characteristic of the subjects studied. They are particularly likely to be misleading or misinterpreted under conditions where nutrient retention is thought to be normal or desirable. The accuracy of any balance study is unknown and relatively small errors would shift the data from a positive to a negative balance or vice versa. When the calculated N gains or losses are small, these may persist for very long periods without appreciable implications in terms of body weight or function. If there is a transition in the metabolic pattern of the subjects as intakes change, this will presumably be dependent upon the prior treatment of the subject and possibly many other things and a typical balance period of 10 days to 2 weeks may not characterize the situation accurately.

Diet type and alteration

It is generally accepted that balancing carbohydrate and protein in the ration has potential impact on N loss and NUE. Further, balancing ruminally undegraded (RUP) and degraded protein (RDP) and use of protected methionine along with strategic selection of protein supplements that are relatively rich in lysine, may permit a 10 to 15% reduction in total N excretion, with most of the reduction occurring in urinary N (Satter *et al.*, 2002). Shriver *et al.* (2003a) used a test diet high in synthetic amino acids in the feeding of pigs. This diet resulted in a drop in urine pH of 0.4 pH units. Ammonia emission is very sensitive to pH, and thus apparent N retention would be affected simply as a result of the change in urine pH due to diet type and alteration. Fiber addition to the diet may result in a shift of N excretion from urine to feces because microbes in the large intestine, thriving on fiber, accrete N that comes from urea in the blood stream, originally destined for excretion in the urine (Shriver *et al.*, 2003b). Lowering urinary N excretion, however, can be expected to affect ammonia losses from urine, in turn affecting apparent N retention.

Animal status

In any experimental situations, a healthy animal is considered to behave normally, while a depleted or diseased (pathology) animals will compound errors of intake, digestibility and nutrient utilization. Owen (1967) demonstrated that nitrogen depletion prior to balance experiments resulted in much higher balances than were observed in adequately fed animals. The balance data for the depleted animals would have supported estimations of unrealistic growth and (or) production figures. Augmented retention during nutrient repletion would be expected. However, it would also be expected that when the deficit was resolved, nutrient balance would drop. It is not, therefore, valid to extrapolate balances obtained during trials to long-term growth or retention figures. This raises the question of the nature of the adaptation period for a balance study. It is obvious that such a period is needed, especially if the steady state assumption is being used for excreta collection. The necessary length of such a period is a matter of some controversy.

Adaptation period in metabolism stalls/cages

Animals need to be adapted to the new environment of metabolism cages having arrangements for the collection of faeces and urine. Some of the animals may behave abnormally in the metabolism cages and a relative stress factors do exist for all types of animals. They may show excitation, refuse intake or fall sick. In that case the more susceptible animals needs to be replaced and not compromised, which often occurs accounting to misleading results and interpretation. Therefore, a steady-state must also be achieved prior to actual collection of faeces and urine.

Losses other than faeces and urine

Calloway *et al.* (1971) have studied exhaustively all possible N losses in man and concluded that about 0.5 g per day of nitrogen is lost through means other than faeces and urine. In ruminant species, gaseous ammonia losses could be appreciable, which is not generally accounted for. Then, possible losses due to fermentation, segregation, sedimentation and volatilization thus exaggerate the overestimation of feed and underestimation of faecal and urinary losses. All the aforementioned discrepancies of determination suggest that balance studies are valid only as relative comparisons in carefully controlled studies but absolute values should be regarded with some suspicion.

Proportional faecal and urinary N excretion

In general, urine is the predominant contributor to volatilization of NH_3 when feces and urine are compared (Kellems *et al.*, 1979) because urea-N is rapidly converted to NH3 due to prevalence of urease enzyme (Mobley et al., 1995). Diet influences not only the amount of N excreted, but also the relative proportion of N in urine and faeces. This can be important in that most of the urinary N will be in the form of urea, and urea can be quickly hydrolyzed by faecal microbes to ammonia when urine and faeces mix following excretion. Ammonia is very vulnerable to being volatilized and lost to the atmosphere. Comparatively, N in faeces is stable and less likely to be volatilized (Satter *et al.*, 2002). To make a quantitative account, urinary N accounts for 45 to 60% of total N excreted, and faeces 40 to 55% (Knowlton *et al.*, 2001; Tine *et al.*, 2001), but there can be significant deviation from these ranges depending on factors related to feeds and feeding and productive state of animals. Realistically, a 10 to 15% reduction in total dietary N in typical dairy diets is perhaps the maximum we can hope for with application of improved feeding programs, but it will translate into about a 13 to 20% reduction in manure N, and perhaps 30 to 35% reduction in N that is vulnerable to volatilization (Satter *et al.*, 2002). Lowering urinary N excretion, however, can be expected to affect ammonia losses from urine, in turn affecting apparent N retention.

Variability in endogenous N loss

It is also necessary in the design and interpretation of balance experiments to consider the role of endogenous nutrient losses. Although this component is not involved in the gross accounting of input minus output, it still has biological meaning and represents the accretion of the nutrient by the organism, if only transitorily. A net N balance of 0 does *not* mean that there was no N deposited in the animals tissues. It means that the animal retained an amount of N equal to its net endogenous losses. Since both tissue retention and endogenous losses are the result of very dynamic systems, it is not possible to know exactly the disposition of this nitrogen in the tissues. However, over a period of time of nitrogen equilibrium in the healthy adult, this represents the total possible anabolic activity.

The endogenous sources of N include saliva, gastric secretions, sloughed cells, and cell debris originating from the stomach and intestinal epithelium, bile, small intestinal secretions, pancreatic secretions, mucus, microorganisms, and mesenteric arterial blood. There are also secretions from the circulation (urea, ammonia) into the small and upper large intestine. Flows of sloughed colonocytes and colon microbiota into the lumen of the large intestine, although the amounts are less than those from the upper gastrointestinal parts, also represent endogenous N (Fuller and Reeds, 1998). Most of the endogenous N can be reabsorbed before reaching the terminal ileum and, to a much lesser extent, in the large intestine and the process is referred to as N recycling, which is affected most by intestinal health and diet type (Bergen and Wu, 2009). In growing pigs, endogenous N flow leaving the small intestine (determined by collecting digesta at the terminal ileum) represents 33% of dietary N intake and can be substantially increased in response to dietary antinutritional factors, high levels of fiber, and protein intake (Grala *et al.*, 1998).

Duration of sample collection

The duration of sample collection varies with ruminants and non-ruminant species. Since the feed retention time in ruminants extends up to 72-96 h, the adaptation and collection period must be higher than non-ruminants. For ruminants, and usually for all other species, excreta collections are made on the basis of the steady state concept. If the animal eats the same amount every day of the same food, there is no need for faeces and urine to correspond literally with the food. However, the steady state must be patent and this condition is very difficult to achieve. In short-term experiments, even if steady state is achieved, the very practical question of which faeces and urine to include and which to discard is raised. A sample collection period of seven days has limited logic except that it facilitates weekly recording. On the contrary, the void by animals is not constant in day-to-day basis and if one examines the values, it will certainly gives a fair idea that the void is rhythmically, if low on day 1 will be high on day 2. This gives a multiple of 2 for the total collection period, which should be at least 4-10 days in both non-ruminant and ruminant species.

Season of collection

There is variability in volatilization rate in different seasons of the year, which may be attributed to varied atmospheric temperature and humidity that affects ureolytic activity (Mobley et al., 1995) and thus difference in ammonia volatilization. There will be more chances of N being volatilized from the urine collection trays during summer compared to winter collection and thus extra care needs to be adopted in minimizing N loss.

2.2.3. Collection and sampling of faeces and urine

Feed wastage is seldom homogeneous and thus, the finer portion tends to be rejected and is usually higher in some nutrients. In addition, moisture loss, drying losses, dust losses and sedimentation in excreta exaggerate the underestimation of excreta losses. Such errors increase with the level of nutrient in the feed. Another error is the lack of correspondence of the excreta sample to the food consumed, e.g. in simple-stomached species, the use of faecal markers can, to some extent, assure that the faeces collected correspond to the food consumed; but there is no way to insure that the urine collected does the same. Further, a thin bacterial film is generally observed in urine collection trays or bottles positioned below the metabolism cages. This film may well be responsible for capturing a substantial portion of urinary N and converting it, or at least some of it, into volatile ammonia.

Preservation and storage

Typically, urine is acidified in the collection container, which is very effective at reducing N losses from the captured urine. Another method is to cool the collection container with ice (often used when urine has to be collected in its native state). In vitro studies have shown that urine contaminated with a small amount of fecal material and stored in a thin layer at room temperature will release 13% of urea-N at 8 h, 26% at 12 h, and 68% at 24 h (van Kempen et al., 2003).

Much of the effort expended on nutrient management has focused on the post-excretion product. It is important to keep in mind that management of the diet can have important impacts on quantitative and qualitative aspects of the excreted nutrients. By reducing dietary N content of dairy diets, total N excretion as well as the proportion of excreted N that is in a form vulnerable to volatilization.can be significantly reduced. Short-term balance experiments which have been used extensively to determine nutrient requirements, especially during growth, amy laed to erroneous conclusions unless the results can be checked with feeding studies which extend over meaningful periods of time. Researchers should take great care to minimize ammonia losses during metabolism trials. Acidification of the entire urine collection system, including the collection tray, is the preferred method, except where urine has to be collected as it is. In those cases, frequent collections and cleaning of the entire urine collection system are important. After analyzing samples, researchers should critically inspect their efficiency figures and determine whether N retention is in line with observed growth of the animals.

The following might be suggested as a checklist for planning or evaluating balance studies:

- What type of trial, balance or retention, is appropriate or possible?
- What is the minimum acceptable technique with regards to precision? (net *vs* gross, apparent *vs* true).
- What care and safeguards have been taken to insure accurate intake and output values?
- Is the trial long enough?
- Are there enough animals and the design is correct?
- Is the steady state achieved prior to actual collection in the metabolism trial?
- Has scrupulous care been taken in planning the level of the test nutrient?
- Has the level of all other nutrients been studied and controlled in keeping with the aims of the experiment?
- What philosophy of adaptation was followed? Was it appropriate?
- Were the data interpreted in the light of the paradox of adaptation philosophy?
- How are samples handled? Is sample aliquot is true representative of the volume excreted?
- Has the samples collected under appropriate conditions and stored taking care of the possible losses?

References

Asplund, J.M. 1979. Interpretation and Significance of Nutrient Balance Experiments. J. Anim. Sci., 49: 826-831.

Bergen, W.G. and Wu, G. 2009. Intestinal nitrogen recycling and utilization in health and disease. J. Nutr.,139:821-825.

Chung, T.K., and Baker, D.H. 1992. Ideal amino acid pattern for 10-kilogram pigs. J. Anim. Sci. 70:3102–3111.

Forbes, G.B. 1973. Another source of error in the metabolic balance method. Nutr. Rev. 31, 297-300.

Fuller M.F. and Reeds P.J. 1998. Nitrogen cycling in the gut. Annu Rev Nutr. 18:385–411.

Galloway, D.H., Odell, A.C.F. and Margen, S. 1971. Sweat and miscellaneous nitrogen losses in human balance studies. J. Nutr. 101, 775-786.

Grala, W, Buraczewska, L, Wasilewko, J, Verstegen, M.W.A., Tamminga, S, Jansman, A.J.M., Huisman, J, Korczynski, W. 1998. Flow of endogenous and exogenous nitrogen in different segments of the small intestine in pigs fed diets with soyabean concentrate, soyabean meal or rapeseed cake. J Anim. Feed Sci. 7:1–20.

Kellems, R., Miner, J.R. and Church. D.C. 1979. Effect of ration on waste composition and length of storage on the volatilization of ammonia, hydrogen sulfide and odors from cattle waste. J. Anim. Sci. 48:436–445.

Mitchell, H.H. and Hamilton, T.S. 1949. The dermal excretion under controlled environmental conditions of nitrogen and minerals in human subjects, with particular reference to calcium and iron. J. Biol. Chem. 178, 345-361.

Mobley, H.C.T., Island, M.D. and Hansinger, R.P. 1995. Molecular biology of microbial ureases. Microbiol. Rev. 59:451–480.

Moorby, J.M. and McConochie, H.R. 2002. A review of nitrogen utilisation efficiency in the dairy cow, and what milk producers can do now to improve it. Department for Environment, Food and Rural Affairs for project WA0322. Institute of Grassland and Environmental Research, Aberystwyth.

NRC. 1998. Nutrient Requirements of Swine. 10th ed. Natl. Acad. Press, Washington, DC.

NRC. 1998. Nutrient Requirements of Dairy Cattle. 7th Revised Edn. Natl. Acad. Press, Washington, DC.

Satter,L.D., Klopfenstein T.J. and Erickson, G.E. 2002. The role of nutrition in reducing nutrient output from ruminants J Anim Sci. 80:E143-E156.

Shriver, J. A., Carter, S. D., Sutton, A.L., Richert, B.T., Senne, B.W. and Pettey, L.A. 2003. Effects of adding fiber sources to reducedcrude protein, amino acid-supplemented diets on nitrogen excretion, growth performance, and carcass traits of fnishing pigs. J. Anim. Sci. 81:492–502.

van Kempen, T.A.T.G., Baker, D.H. and van Heugten, E. 2003. Nitrogen losses in metabolism trials. J. Anim. Sci. 81:2649-2650.

Wallace, W.M. 1959. Nitrogen content of the body and its relation to retention and loss of nitrogen. Fed. Proc. 18:1125–1130.

2.3. Procedures and General Consideration for Nutrient Balance Trials with Ruminants

2.3.1. Selection and care of animals

- Select animals of similar size, sex, age, weight, and breeding. They should be free of parasites, healthy, and vigorous; but not unduly excitable.

- Place animals in preliminary feeding pens, and allow them to become adjusted to environment and to the routines of handling and feeding. Gradually change the diets to those that will be used in the trial. During this adjustment period, replace animals that are "poor eaters", and those that are overly excitable or otherwise unsuitable. Weigh the animals at regular intervals, and keep a record of these "preliminary weights".
- Allow the animals a few days for adjustment to metabolism cages after their transfer from the preliminary holding pens. Protect them from flies and pests, keep them clean, and provide for their comfort within the limits of the trial objectives.
- Keep a daily record of temperature, weather conditions, unusual events in the animal laboratory, and any observations of pertinence to animal behavior or performance.

2.3.2. Experimental feeds

- During the preliminary period, calculate the amount of diets (roughage, concentrates, and supplements) necessary for the entire period of the study. Trials with ruminants should consist of at least a 10-d "adjustment period" followed by a "collection period" of 10 d. It is wise to provide a surplus of all dietary ingredients.
- Ensure that all feed ingredients are uniformly mixed, so that ingredients fed initially are comparable to those fed finally. In the case of concentrate mixtures and supplements, mix amounts sufficient to provide at least 11;4 to 11;2 times the amount calculated to be fed during the actual trial. This allows for "repeat" days, spillage, and so on.
- Store all feed ingredients in covered containers conspicuously labelled and completely labelled :

  i) Project number

  ii) Trial number

  iii) Animal number (to receive diets)

  iv) Dates of trial

  v) Experimenter

  vi) Secure the labels on containers! (Never label lids only). Situate feed cans conveniently, and protect against inadvertent interchange of locations. For roughages, select a quantity of feed that is in excess of needs, uniform in quality, and properly identified. Store conveniently, and protect from contamination especially dusts from feedstuffs and chemicals.

- Preliminary sampling and analyses. Take a representative sample of all feedstuffs and analyze before the onset of the actual trial to ensure that composition of diets is as prescribed for the trial.

Procedures

1. Check the accuracy and precision of all scales, balances, and volumetric equipment to be used in measurement of feedstuffs and excreta.
2. Plan in detail a definite routine to be used in the trial - both for the preparation and provision of diets and for the collection of excreta - and follow this routine throughout. Generally, this involves weighing dietary ingredients directly into individual feed pans, and, if necessary, mixing each animal's diet by hand to ensure uniform distribution of ingredients and to prevent selective consumption by the animal.
3. During the preliminary period, while animals are in metabolism cages, collect daily a "running grab sample" from each of the dietary ingredients. Composite these into single, 10-d samples for each of the diets. Label thoroughly! Use the same procedure to collect additional 10-d, composite samples for each of the ingredients fed during the subsequent 10-d collection period. Label Completely.
4. Feed consumed during the adjustment period will be excreted during the subsequent collection period. Therefore, the daily feed must be kept constant throughout the 20-d feeding period - constant in amount, and unvarying in composition.
5. Maintain a "Bam Record" showing the daily feeding performance of each animal, the daily excreta, and any notes of consequence.
6. At the onset of the "collection period", start the collection of urine and fecal samples at a specified time, and strive to meet the same schedule throughout the trial. Be sure to end the trial at the same time on the final day of collection.
7. If there should be any feed uneaten from feeding time to feeding time, remove the entire refused portion, weigh and record the

weight, and collect a *representative* sample for analysis. Lable Completely.

8. *Collection of urine*: Equipment used must afford quantitative collection of entire 24-h excreta.
   i) Collect urine into clean containers that are safe-guarded against contamination from feedstuffs, fecal material, and extraneous material (e.g., flies, bugs, dust, and so on). Protect also from spillage and breakage (plastic jugs are preferred in most cases).
   ii) Add a few drops of toluene into the collection jugs at the onset of each 24-h period to retard microbial activity. (this will have insignificant effects on most analyses that are conducted in routine trials).
   iii) When making the daily collections, provide a temporary container to collect any excretion made during the process (a "stitch in time ... ").
   iv) During the collection, suspend sediment by shaking the collection jug, and pour quickly into a volumetric cylinder (graduated), and record the daily urine volume to ± I mL. Test the reaction of the urine to litmus (note if basic) and determine specific gravity and(or) total solids by refractometry, if applicable.
   v) Dilute the urine to a prescribed, constant volume (generally 1 L) with distilled water, mix thoroughly by pouring back and forth between two containers, and collect a prescribed portion (generally 100 mL per day from a l-L total volume) into the container for the 10-d composite sample. Make sure that each of the 10 daily collections is represented equally in the 10-d composite sample. (It is wise to make duplicate, 10-d collections for each animal. Use of plastic bottles for these composite samples will both allow freezing, if desired, and also prevent breakage.)
   vi) Add to the collected portion of urine an amount of concentrated HCl to ensure that the sample is slightly acid to litmus (avoid excess). Check reaction to litmus each day, and add acid only when necessary. RECORD the amounts of acid added. Add a few drops of toluene to the composite sample, (as well as to the collection jugs), stopper the composite samples, and store under refrigeration (approx. 4°C).

vii) *Label Completely :*

a) Project number

b) Trial number

c) Animal number

d) Dates of collection

e) Experimenter

f) Inspect the apparatus and assembly for urine collection daily (or more often) to ensure that they are clean and free-flowing.

9. *Collection of feces*: Feces must be collected daily. Follow a prescribed routine.

   i) Weigh the entire excretion. Mix thoroughly, and collect a *representative* sample amounting to one-tenth of the daily excretion. Record the weights of daily excretions for each animal.

   ii) Into the container for the composite sample of feces add a few thymol crystals (approx .. 1 gram per 100 grams of feces collected into the vessel); mix, seal, and store under refrigeration.

2.4. Estimation of Rumen Microbial Protein Production from Purine Derivatives in Urine

Principle

Nucleic acids leaving the rumen are mainly of microbial origin which are extensively digested in the small intestine and the purines liberated are absorbed. Only a small amount of absorbed purines are utilised by the host animal, the rest are metabolised forming hypoxanthine, xanthine, uric acid and allantoin. These metabolites are excreted mainly through urine. Therefore, with the understanding of how urinary excretion of purine derivatives (PD) respond to purine absorption i.e. the response curve of PD excretion to purine input into intestines, the microbial purine outflow from the rumen and the microbial N supply to the ruminant animal can be estimated by measuring the purine derivatives excretion through urine.

Sample collection

1. Collect all the urine produced by the animal without mixing of faeces.

2. Urine collection should be made for more than 5 days to reduce the error due to the end of collection variation in urine output.
3. Collection can be made as a bulk for the whole period.
4. Urine is collected into a container with approximately 100 ml. of 10% H2S04.
5. Record the original weight of the urine. The daily urine output may be 0.5-2 lit in sheep and 20-30 lit. in cattle. Add tap water to constant weight (4Kg for sheep, 50 kg or more for cattle). So the final volume of diluted urine is the same for all animals and every day.
6. Mix the diluted urine thoroughly, filter some urine through glass wool or surgical gauze, take a subsample of 20 ml and store it at - 20°C.

Dilution of urine samples

The urine samples which have been previously diluted before storage still needs further dilution. The next dilution should be to such an extent that the concentrations in the final samples will be within the range of the standards used in the assays (5-50 mg/lit for both uric acid and allantoin).

Determination of purine derivatives

Determination of allantoin by colorimetric method

Principle

Allantoin is first hydrolysed under a weak alkaline condition at 100°C, to allantoic acid which is further degraded to urea and glyoxylic acid in weak acid solution. The Glyoxylic acid then reacts with phenylhydrazine hydrochloride to produce a phenylhydrazone of the acid. The product then forms an unstable chromosphore with potassium ferricyanide. The colour is read at 522 nm.

Preparation of standards

(a) Prepare an allantoin *stock solution* of 100mg/lit. Weigh 50 mg of allantoin and transfer it into a volumetric flask. Dissolve in about 100 ml 0.01 M NaOH and make up the volume with distilled water.

(b) Prepare *working concentrations* of 10, 20, 30, 40, 50 and 60 mg/l. (c) Store each working standards as small aliquots in the freezer.

Procedure

1. Pipette 1 ml of sample, standard or distilled water (blank) into 15 ml tubes
2. Add 5 ml of distilled water.
3. Add 1 ml of 0.5M NaOH.
4. Mix the contents of the tubes by Vortexing.
5. Put the tubes in the boiling water bath for 7 min.
6. Remove from the boiling water and cool the tubes in the cold water.
7. Add to each tube 1 ml of HCI (0.5M). The pH after adding HCl must be in the range of 2-3.
8. Add 1 ml of phenylhydrazine solution (0.023 M freshly prepared). Mix and transfer the tubes again to the boiling water both for exactly 7 min.
9. Remove from the boiling water and dump it immediately mto the icy alcohol bath for several minutes.
10. Pipette 3 ml of cone. HCI (11.4N) (operate in a fume cupboard) and 1 ml of Potassium ferricyanide (0.05 M). Prepare K-ferricyanide freshly on the day of testing. Perform this for ali samples within a shortest possible span. Start the timer.
11. Mix thoroughly and transfer some to 4.5 ml cuvettes at room temperature.
12. Read the absorbance at 522 nm after exactly 20 min. Once started, do it as quickly as possible.

Calculation

The standard curve is linear. Therefore. we can fit a linear regression between the known allantoin concentrations (Standards) (X) and the corresponding Ol) (Y). The slope of the line is usually 0.16-0.18. Calculate the concentration of the unknown based on this equation.

Determination of allantoin by HPLC

Reagents

1. 1.0 giL 2,4-dinitrophenylhydrazine (DNPH) dissolved In 2M HCI and filtered through Whatman No.1 paper. Store at 4°C.

2. 0.6M NaOH
3. pH indicator: 0.04 (w/v) thymol blue, dissolved in distilled deionized water and filtered through Whatman No.1 paper.
4. Allantoin standards: 5, 10, 20, 30, 40, 50 mgll water.

Derivatization procedure

1. Do a set of standard for each batch of derivatization.
2. Pipette 500lli of a sample or standard to a 1.5 ml microcentrifuge tube.
3. Add 50 III of the pH indicator. The colour of the mixture would be either yellow or red (red when pH<1.2).
4. Add 100III of 0.6M NaOH solution.
5. Mix and observe the colour. If the colour fails to change to blue (pH>9.2
6. when blue), add 50 III,NaOH (volume to be recorded).
7. Repeat (e) if necessary
8. Cap tubes, mix. Place them in water bath at 85°C for 1 hour.
9. Transfer the tubes from the water bath, add 200 III of the DNPH solution.
10. Mix. the colour of the mixture that would be orange-yellow.
11. Place the tubes back into the water bath to continue heating for another 20 min.
12. If the samples are not used for injection into HPLC right away, store the sample in the freezer.
13. Centrifuge the samples at 20,000 rpm for 25 min.
14. Carefully transfer 0.3 ml. or 0.5 ml of the supernatant into each sample vial.
15. Make sure that the vials are clearly labelled and there are no air bubbles in the vials.
16. These samples are now ready for injection. Store in the fridge if necessary.

Calculations

Calculate the concentrations of unknown with the known standards.

Determination of xanthine plus hypoxanthine by enzymatic method

Principle

In this method, xanthine and hypoxanthine are enzymatically converted to uric acid and that uric acid is monitored by its absorbance at 293 nm.When the urine samples are treated with xanthine oxidase, xanthine and hypoxanthine are converted to uric acid. There should be an increase i;:} OD at 293nm after the enzyme treatment. The net increase is then used for the calculation of the amount of uric acid formed based on the uric acid standard curve.

Standards preparation

Prepare a stock uric acid solution of 100 mg/l. Dilute it to give working concentrations of 20, 40, 60, 80, 100mg/l.

Store each working standard as small quantities in the freezer.

Procedure

1. Pipette 1 ml of urine or standard or distilled water into a cuvette. All samples and standards are done in duplicates. Distilled water is used as the blank. Prepare two sets.
2. Add 2.5 ml phosphate buffer(KH2P04 0.2 M, pH 7.35 adjust with either HaP04 or KOH)
3. Add 0.35 ml of 4.3 mML-histidine solution and Mix.
4. In one set, add 250 III of buffer, in the other set add 150 III of the xanthine oxidase solution (xanth ine oxidase 25 ul to 3 ml of the buffer).
5. Mix well and incubate at 37°C for 60 mins.
6. Read OD at 293 nm.

Calculations

Use OD of the standards without XO added for construction of uric acid standard curve. Transform both X and Y into natural logarithmic function. Fit the Ln 00 into a linear function of Ln (X).

Calculate the OD for the samples, i.e. the difference between two sets with and without XO addition:

OD= OD with XO - OD without XO.

Calculate the corresponding concentration of uric acid from OD based on the above standard curve.

Estimate the contribution of OD reading from the xanthine (ODX) in the set without XO based on a predetermined xanthine standard curve.

Re-adjust the OD (i.e. OD2 = OD + ODX) and repeat from step 3 once.

The uric acid standard curve is not linear, but when both X (concentration) and Y (OD) are transformed into Ln(X) and Ln 00, LN 00 becomes a linear function of Ln(X). This relationship is then used for the calculation of the concentrations of the samples from their OD readings.

Determination of uric acid by uricase method

Principle

Uric acid absorbs UV at 293 nm, although other compounds may also absorb at this wavelength. When urine samples are treated with uricase, uric acid is degraded into allantoin and other compounds that do not absorb UV at 293nm. Therefore, the reduction in OD reading after treatment with uricase is related with the concentration of uric acid in the sample. After treatment, OD of the standards should be zero if the conversion is complete.

Standards preparation

As in the case of xanthine. In this case also, prepare standard working concentration of 5, 10, 20, 30 and 40 mg/l.

Enzyme preparation

Dilute the enzyme in the buffer to obtain a concentration of 0.12 Ulml (Uricase from Porcine liver (ex. Sigma Cat No. U-9375, 19 unitslg solid). Keep the enzyme solution in the freeze to preserve the enzyme activity.

Procedure

1. Two sets of standards and a 'blank' (using distilled water) and samples in duplicate, are prepared.

2. Pipette 2.5 ml of urine or standard or water into 10 ml tubes.
3. Add 1 ml phosphate buffer (KH2P04, 0.67M, pH 9.4 adjust the pH with KOH).
4. Mix well the contents of the tubes by Vortexing.
5. In one set, add 150 III buffer and in the other add 150 III of uricase solution
6. f) fix again by vortexing and incubate in the water bath and 37°C for 90 min.
7. Remove from water bath, mix and transfer again the solutions to the cuvettes and read the OD at 293 nm. If the enzymatic conversion is complete, the OD of the standards with uricase added should be zero. If not, incubate in water bath for an additional 30 min. and read again.

Standard curve and calculations

Standard curve is curvilinear. When both X and Yare transformed to Ln functions, Ln (Y) is linearly correlated to Ln (X).

Use the OD reading of the set without addition of uricase for the construction of standard curve.

Calculate the net reduction in OD for the samples due to uricase treatment.

Actual OD = OD without enzyme - OD with enzyme.

Calculate the uric acid concentration from OD based on the established standard equation.

Calculations for daily excretion of purine derivatives

Calculate the excretion of allantoin, uric acid and xanthine plus hypoxanthine, unit of m mol/dl.

The total PD excretion (i.e. sum of all compounds) is used for the estimation of microbial protein supply.

In sheep the sum of all 4 compounds and in cattle the sum of allantoin and uric acid are taken into consideration. The proportion of the individual components expressed as % of the sum are approximately as follows.

In sheep = allantoin 60-80%; uric acid 30-10%, xanthin plus hypoxanthine 105%. As the total excretion increases, the proportion of allantoin increases.

In cattle: Allantoin 80-85%, uric acid 20-15% within the same animal, the proportions are very constant, but there seems to be variation between animals.

Calculation of microbial N supply

(A) Purine absorption and PD excretion

Different equations are used for sheep and cattle to describe the quantitative relationship between absorption of micrbial purines (X mmol/dl), and excretion of PD in urine(y mmol/dl):

Sheep $Y = 0.84\ X + (0.150\ W^{O.75}\ e\text{-}0.25X)$ (1)

Cattle $Y = 0.85\ X + (0.385\ WO.75)$ (2)

Where $W^{0.75}$ represents the metabolic body weight (kg) of the animal.

The slopes of 0.84 and 0.85 in Equations (1) and (2) respectively, represent the recovery of absorbed purines as PD in urine. The component within parenthesis represents the net endogenous contribution of PD to total excretion after correction for the utilisation of microbial purines by the animal. In cattle, the endogenous contribution is taken as a constant at 0.385 mmol/kg $W^{0.75}$ per day. In sheep, this reduces to zero as exogenous purines are utilised and de novo synthesis of purines is phased out.

(B) Calculation of daily purine absorption

Based on equations (1) and (2) above for sheep and cattle, respectively the amount of exogenous purines absorbed can then be estimated from the daily excretion of PD.

Sheep

The calculation of X from Y based on equation (1) can be performed by means of the Newton-Raphson iteraction process and we finally arrive at equation:

Where $f\ (x) = 0.84\ X + 0.150\ WO.75\ e\ \text{-}0.\ 25x\text{-}y$

And the derivatives of $f(x) = 0.84 - 0.038\ W^{075}\ e - 0.25x$

Example = A sheep of 60 kg CVV) excreted 5.0 mmolJd PD 00. We give an initial value Xi = Y +0.84 = to put in equation (3) to calculate X2 Now we obtain X2 = 4.838 Next we put X2 back to equation (3) to calculate X3 (X3 = 4.790), and so on Xa and X5 can be calculated as 4.790 and 4.790, respectively. We will notice that, as the iteraction process goes on, Xn approaches a constant value, which is the answer. In this example, the amount of microbial purines absorbed is therefore 4.79 mmolJdl.

Cattle

The calculation is straight forward

X= (Y-0.385 X W075) 10.85 (4)

Example: A bull of 321 kg live weight, the daily excretion of purine derivatives was 153 mmol/dl. The amount of microbial purines absorbed can be calculated as:

(153-0.385 x 321 0.75)/0.85=145.7mmol/d.

(C) Calculation of intestinal flow of microbial N

The following factors are used for the calculation of intestinal flow of microbial N (gN/d) from the microbial purines absorbed (X mmolJdl) (Equation5).

(i) Digestibility of microbial purines is assumed to be 0.83.

(ii) This is taken as the mean digestibility value for microbial nucleic acids based on observations reported in the literature.

(iii) The N content of purines is 70 mgN/mmol.

(iv) The ratio of purine N: total N in mixed rumen microbes is taken as 11.6:100.

Microbial N(gN/d) = X (mmolJdl) x 70 0.116 X 0.83 X 1000

The results of the calculated microbial protein supply are expressed as "g' microbial N per day" and/or as "g microbial N per kg digestible organic matter intake (DOMI)". The latter is effectively an expression of the efficiency of microbial protein supply. The efficiency can also be expressed as "g' microbial N per kg digestible organic matter apparently fermented in the rumen (DOMR). DOMR can be assumed to be 0.65 of the DOMI.

Reference

LA.E.A. 1997. Estimation of rumen microbial protein production from purine derivatives in urine. A laboratory manual, IAEA - TECDOC - 945. IAF A, Vienna.

□□□

Chapter 3

In vitro Feed Evaluation

A major constraint to livestock production in developing countries is the scarcity and fluctuating quantity and quality of the year-round feed supply. Providing adequate good-quality feed to livestock to raise and maintain their productivity is and will be a major challenge to agricultural scientists and policy makers all over the world. Future hopes of feeding the millions and safeguarding their food security will depend on the enhanced and efficient utilisation of unconventional resources that cannot be used as food for humans, as feed for livestock. In developing countries, livestock are fed mainly on agro-industrial by-products containing a larger proportion of ligno-cellulosic feeds like cereal straws, stovers, sugarcane by-products and similar other feeds. These feeds are poor in protein, energy, minerals and vitamins. The use of simple but robust techniques for evaluation of the nutritional quality of these feed resources will contribute to their efficient utilization. The determination of intake and digestibility of feedstuffs *in vivo* is time-consuming, laborious, expensive, requires large quantities of feed and is unsuitable for large-scale feed evaluation. Therefore, many attempts have been made to predict intake and digestibility using laboratory techniques. Much effort has been directed towards the development of regression equations to predict digestibility from forage chemical composition, but a regression equation that satisfactorily predicts a wide range of forages has not yet been derived. This paper highlights the potential of a novel approach using an *in vitro* rumen fermentation technique for evaluation of the nutritional quality of conventional and unconventional feed resources.

The mission for a simple, cheap and reliable laboratory method to evaluate the nutritive value of feed resources for ruminant animals was a very difficult task for the animal nutrition scientist and it resembles the search for the philosopher's stone. It was an urgent need for such a technique, not only are cost and animal-welfare considerations making the use of animals less desirable, but also the increasing human population and demand for animal products create a need for new animal feedstuffs and improved varieties of traditional ones. The microbial degradation of forages in the rumen is a complex process, involving complex interactions, both between the microorganisms present (bacteria, protozoa and fungi), and between the microbial population and the host. To develop such a method, the findings of rumen microbiologists published for the last 40 years provide an important source of information. For example, factors that must be considered when trying to simulate the rumen environment include temperature, pH, buffering capacity, anaerobiosis and nitrogen source. Existing laboratory techniques concentrate on estimating rumen degradation. The rationale for this is that forage consists largely of carbohydrate in the form of fibre, and this can only be digested by microorganisms in the rumen. However, the effective degradability (the digestibility that can be expected *in vivo*) also depends on the rate of degradation, the potential degradability and the fractional outflow rate from the rumen . At the moment, there is no *in vitro* technique that can estimate the lattermost aspect of *in vivo* digestibility. Existing techniques measure the potential degradability or both the degradability and rate of degradation (*in situ*, and cumulative gas production).

The *in situ* technique has been widely used for the last 50 years. Whatever it is called (*in sacco*, nylon bag, Dacron bag, cloth bag), it requires the incubation of substrates in bags within the rumen of fistulated animals. These bags are removed at regular intervals, the weight loss is plotted (usually as dry matter (DM)) and a curve is fitted to the resultant data points. Van Soest (1994) has listed the advantages and disadvantages of this kind of technique. While by no means a new concept, the measurement of gas production to evaluate forages has come to prominence in the last 10 years. There are several methodologies in current use, and the differences and similarities between them will be described, along with current research into the fundamentals of what the technique can predict and how best to carry it out. The general stoichiometry of gas related to volatile fatty acid (VFA) production is described for the rumen fermentation process. The use to which the technique is being put and potential future developments are then discussed in relation to feed evaluation.

In vitro Methods

In vitro methods for laboratory estimations of degraded feeds are important for ruminant nutritionists. An efficient laboratory method should be reproducible and should correlate well with actually measured *in vivo* parameters. *In vitro* methods have the advantage not only of being less expensive and less time-consuming, but they allow maintaining experimental conditions more precisely than do *in vivo* trials. Three major biological digestion techniques are currently available to determine the nutritive value of ruminant feeds: 1) digestion with rumen microorganisms) or using a gas method, 2) *in situ* incubation of samples in nylon bags in the rumen, and 3) digestion with cell-free fungal cellulases. These biological methods are more meaningful since microorganisms and enzymes are more sensitive to factors influencing the rate and extent of digestion than are chemical methods. The nylon bag technique has been used for many years to provide estimates of both the rate and extent of disappearance of feed constituents. This technique provides a useful means to estimate rates of disappearance and potential ruminal degradability of feedstuffs and feed constituents whilst incorporating effects of particulate passage rate from the rumen. The disadvantage of the method is that only a small number of forage samples can be assessed at any one time, and it also requires at least three fistulated animals to account for variations due to animal. It is therefore of limited value in laboratories undertaking routine screening of a large numbers of samples. It is also laborious, and requires large amounts of samples. Substantial error could result in values obtained at early stages of digestion due to a low weight loss; and for poor quality roughages, adherence of microbes at early stages can even lead to higher weights and thus distortion of results if kinetic modelling does not incorporate the lag.

The Tilley and Terry technique is used widely because of its convenience, particularly when large-scale testing of feedstuffs is required. This method is employed in many forage evaluation laboratories and involves two stages in which forages are subjected to 48 h fermentation in a buffer solution containing rumen fluid, followed by 48 h of digestion with pepsin in an acid solution. The method was modified by Goering and Van Soest (1970), in that the residue after 48 h incubation was treated with neutral detergent solution to estimate true dry matter digestibility. Although the method of Tilley and Terry (1963) has been extensively validated with *in vivo* values (Van Soest, 1994), the method appears to have several disadvantages. The method is an end-point measurement (gives only one observation) thus, unless lengthy and labour-intensive time-course studies are made, the technique does not provide information on the kinetics of forage digestion; the residue determination destroys the sample and therefore a large number of replicates are needed. The method is therefore difficult to apply to materials such as tissue culture samples or cell-wall fractions.

In vitro fermentation methods

In vitro digestibility techniques provide a quick, inexpensive, and precise prediction of in vivo or conventionally determined digestibility in ruminants. Essentially, the method simulates the processes that occur in the rumen, and probably the most commonly used in vitro technique is the one devised by Tilley and Terry (1963). Estimates of digestibility by the Tilley and Terry procedure are within 1 to 3% of conventionally determined values. The in vitro procedure does a better job of prediction than chemical composition because it accounts for all factors affecting digestibility, whether known or unknown, which is not possible with current chemical methods. This accounting for additional factors is primarily a function of the use of ruminal fluid from a donor steer as the digestive agent, thereby including unknown factors that simple chemical analyses of the feed do not reveal. As indicated previously, the in vitro procedure is quite simple, but nonetheless subject to a number of variables that may influence the results obtained.

In vitro dry matter disappearance is determined by the following formula:

IVDMD % = 100 × [(initial dry sample wt - (residue – blank))/ initial dry sample wt]

The procedure is useful in that estimates of digestibility can be obtained in a few days on a large number of samples. However, some of the variables that influence the procedure can be grouped under four major heads as detailed below:.

A. Variations in the microbial population
   1. Diet of donor animal
   2. Animal to animal differences
   3. Inoculum processing
B. Variations resulting from different storage, grinding, and processing techniques in sample preparation
C. Differences attributable to the fermentation medium
   1. Sample:inoculum ratio
   2. Buffer
   3. Nutrients in medium
D. Procedural variations such as length of fermentation and laboratory errors.

The largest source of variation among the four major sources listed above is the variation in the microbial population. Difficulties associated with this source of variation can be partially overcome by using more than one animal and by feeding donor animals the same or a similar diet to that being evaluated in the in vitro system. In addition, fluid should be removed at a standard time after feeding. The author's personal preference is to remove fluid ~4 h after feeding when microbial numbers are maximal; however, several researchers prefer removing fluid after the donor has been withheld from feed and water for 12 to 14 h. Generally, the standard method of processing the inoculum is simply to strain whole ruminal contents through at least four layers of cheesecloth.

With regard to variations in the sample, it has been observed that finely ground samples are more highly digested than are coarsely ground samples. Hence, all samples should be ground in the same manner; grinding to pass a 1- 2 mm screen usually is adequate. A commonly used sample:inoculum:buffer ratio is 0.5 g:8 mL of ruminal fluid:32 mL McDougall's buffer. Some laboratories add nutrients to the buffer, the most common being urea, to prevent nutrient deficiencies during fermentation. Fermentation should be maintained at a pH of 6.9 to 7.1 for optimum results. Procedural variations can be minimized by standardizing temperature, time of fermentation, centrifugation speeds, and so on. In fact, the key to successful in vitro analysis is to standardize as much as possible. In this regard, it is a good practice to include a standard forage with each in vitro run as a means of determining the validity of individual runs. All things considered, the in vitro digestibility technique is the best means of laboratory evaluation of digestibility available today. The procedure will continue to be used extensively for some time to come. Both the Tilley and Terry and nylon bag methods are based on residue determinations and may result in overestimation of dry matter digestibilities for tannin-rich feeds, since tannins are solubilised in both these systems but might be indigestible and do not contribute to nutrient supply to animals.

The *In vitro* gas measuring technique has been widely used for evaluation of nutritive value of feeds. The increased interest in the efficient utilisation of roughage diets has led to an increase in the use of this technique due to the advantage in studying fermentation kinetics. Gas measurement provides a useful data on digestion kinetics of both soluble and insoluble fractions of feedstuffs. Several gas measuring techniques and *in vitro* gas methods are in use by several groups. The *in vitro* gas method based on syringes appears to be the most suitable for use in developing countries. Other *in vitro* methods such as Tilley and Terry and nylon bag methods are based on gravimetric measurements which follow disappearance of the substrate (the components which may or may not necessarily contribute to fermentation), whereas gas measurement focuses on the appearances of fermentation products (soluble but not fermentable products do not contribute to gas production). In the gas method, kinetics of fermentation can be studied on a single sample and therefore a relatively small amount of sample is required or a larger number of samples can be evaluated at a time. The *in vitro* gas method is more efficient than the *in sacco* method in evaluating the effects of tannins or other anti-nutritive factors. In the *in sacco* method these factors are diluted in the rumen after getting released from the nylon bag and therefore do not affect rumen fermentation appreciably. In addition, the *in vitro* gas method can better monitor nutrient-antinutrient and antinutrient-antinutrient interactions.

In vitro gas method

A simple *in vitro* approach is described below which is convenient and fast, and allows a large number of samples to be handled at a time. It is based on the quantification of substrate degraded or microbial protein produced using internal or external markers, and of gas or short chain fatty acid (SCFA) production in an *in vitro* rumen fermentation system, most suitably in glass syringes. Fermentations are conducted in 100 ml capacity calibrated glass syringes containing feedstuff and a buffered rumen fluid. The gas produced on incubation of 200 mg feed dry matter after 24 h of incubation together with the levels of other chemical constituents are used to predict digestibility of organic matter determined *in vivo* and metabolizable energy.

For roughages, the relationships are:

ME (MJ / Kg DM) = 2.20 + 0.136 Gp + 0.057 CP, R^2= 0.94

OMD (%) = 14.88 + 0.889 Gp + 0.45 CP + 0.0651 Ash, R^2=0.92

Where ME is the metabolisable energy; OMD organic matter digestibility; CP, crude protein in %; ash in %; and Gp, the net gas production in ml from 200 mg dry sample after 24 h of incubation and after correction for the day-to-day variation in the activity of rumen liquor using a standard feed substrate.

An increase in amount of sample from 200 to 500 mg reduces the inherent error associated with gravimetric determination needed to determine concomitant *in vitro* apparent and true digestibility. In the 30 ml system, the linearity between the amount of substrate incubated and the amount of gas produced is lost when the gas volume exceeds 90 ml because of the exhaustion of buffer of the medium; and in 40 ml system, the linearity is lost when the gas volume exceeds 130 ml. The exhaustion of the buffer decreases pH of the incubation medium; consequently the fermentation is inhibited. If the amount of gas production exceeds 90 ml using the 30 ml system and 130 ml using the 40 ml system, the amount of feed being incubated should be reduced.

When a feedstuff is incubated with buffered rumen fluid *in vitro*, the carbohydrates are fermented to produce short chain fatty acids (SCFA), gases and microbial cells. Gas production is basically the result of fermentation of carbohydrates to acetate, propionate and butyrate. Gas production from protein fermentation is relatively small as compared to carbohydrate fermentation and the contribution of fat to gas production is negligible. The gas produced in the gas technique is the direct gas produced as a result of fermentation and the indirect gas produced from the buffering of SCFA. For roughages, when bicarbonate buffer is used, about 50% of the total gas is generated from buffering of the SCFA and the rest is evolved directly from fermentation. At very high molar propionate the amount of CO_2 generated from buffering of SCFA is about 60% of total gas production. Each mmol of SCFA produced from fermentation releases 0.8-1.0 mmol of CO_2 from the buffered rumen fluid solution, depending on the amount of phosphate buffer present. Highly significant correlation has been observed between SCFA and gas production (see below).

Gas is produced mainly when substrate is fermented to acetate and butyrate. Substrate fermentation to propionate yields gas only from buffering of the acid and, therefore, relatively lower gas production is associated with propionate production. The gas that is released with the generation of propionate is only the indirect gas produced from buffering. The molar proportion of different SCFA produced is dependent on the type of substrate. Therefore, the molar ratio of acetate of propionate has been used to evaluate substrate related differences. Rapidly fermentable carbohydrates yield relatively higher propionate as compared to acetate, and the reverse takes place when slowly fermentable carbohydrates are incubated. Many workers reported more propionate and thus lower acetate to propionate ratio in the ruminal fluid of cows fed a high grain diet. If fermentation of feeds leads to a higher proportion of acetate, there will be a concomitant increase in gas production compared with a feed with a higher proportion of propionate. In other words, a shift in the proportion of SCFA will be reflected by changes in gas production. The stoichiometric relationship between gas production and SCFA is as follows:

$$\text{Fermentative } CO_2 = A/2+P/4+1.5B$$

$$\text{Fermentative } CH_4 = (A+2B)-CO_2$$

where A, P and B are moles of acetate, propionate, and butyrate respectively

Assumption : one mole of SCFA releases one mole of CO_2 from bicarbonate-based buffer described as buffering CO_2 and therefore, mmol of buffering CO_2 is equal to mmol of total SCFA generated during incubation.

The SCFA production could be predicted from gas values using the above relationship. The level of SCFA is an indicator of energy availability to the animal. Since SCFA measurement is important for relating feed composition to production parameters and to net energy values of diets, prediction of SCFA from *in vitro* gas measurement will be increasingly important in developing countries where laboratories are seldom equipped with modern equipments to measure SCFA. The stoichiometric balance also allows calculation of the CH_4 and CO_2 expected from the rumen fermentation if the molar proportions and amount of SCFA are known.

Evaluation of Feed Resources

Recent advances in ration balancing include manipulation of feed to increase the quantity and quality of protein and energy delivered to the small intestine. The limited supply of protein post-ruminally under most feeding systems in developing countries is an important limiting factor that prevents enhancement in animal productivity. Selection of fibrous feeds based on high efficiency of microbial protein synthesis in the rumen along with high dry matter digestibility, and development of feeding strategies based on high efficiency as well as high microbial protein synthesis in the rumen will lead to higher utilization of protein as well as energy. This concept of feed evaluation has an extra element of the efficiency of microbial protein synthesis in addition to the more conventional one of the dry matter digestibility.

3.1. Gas-Production Techniques for Forage Evaluation

3.1.1. Early techniques

Gas (carbon dioxide (CO2) and methane (CH4)) is a waste product of fermentation. In 1943, Quin incubated feedstuffs in gas-tight flasks and measured gas production manometrically. By the 1960s, the measurement of gas using liquid displacement systems was suggested as a technique to evaluate ruminant feedstuffs and this is still used today. In order to simplify measurement of gas by manometry, some workers have used different techniques, whereby gas production was measured by use of a syringe attached to a flask in which the substrate was fermented. They found characteristic patterns of change in rates of gas production according to the substrate used.

3.1.2. The Hohenheim gas test

The group at Hohenheim University (Germany) were the first to conduct the fermentation process within large glass syringes, so that the gas produced pushed the plunger outwards, and the total volume produced was recorded after 24 h. This evaluation system combines the total gas volume after 24 h with the concentration of crude protein (CP), crude fat, crude fibre and ash of the feed to predict metabolizable energy (ME). In Germany, the Hohenheim *in vitro* gas test is now widely used for the estimation of *in vivo* digestibility and ME for ruminants. Recently, some workers have adapted the Hohenheim gas test by recording the change in gas production in the glass syringes at regular time intervals, so that a measure of the kinetics of fermentation is obtained, as well as the end-point value.

Fig. 1 : Equipment for the Hohenheim gas test (photo courtesy of Klaus Aiple and Klaus Becker, University of Hohenheim).

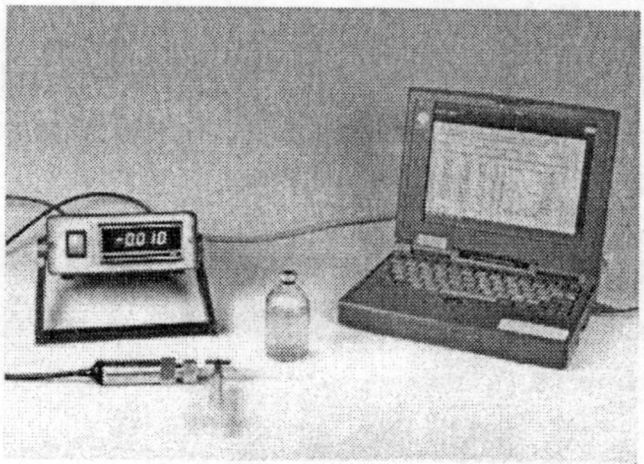

Fig. 2 : The pressure transducer, complete with serum bottle, connectors and computer, as used for the manual transducer technique.

3.1.3. Manual transducer technique

A new method, using a pressure transducer to measure head-space gases in sealed serum bottles, was first described in 1991 from the Institute of Grassland and Environmental Research (IGER) in the UK The pressure and volume were recorded at regular intervals: approximately every 3–4 h for the first 24 h, and then less frequently to the end of fermentation at about 120 or 144 h. The head-space gas is removed after each reading. The cumulative gas profile represents the kinetics of the fermentation process of individual feedstuffs.

Automated systems

Automated system was based on the weight of fluid displaced by fermentation gas for 24 individual bottles. The change in weight was recorded and a calculation made to convert the weight to gas volume, which was then registered by a data logger. Cornell University in the USA, first described the use of computerized pressure sensors to monitor gas production. They investigated the use of several variables, including sample size, inoculum size, vessel size and type of pressure sensor, and reported the ranges within which gas production could be accurately measured. For the Cornell technique, gas is not vented during the fermentation but the head space of the bottles is much larger than for the IGER technique and the sample much smaller. An automated version of the manual transducer technique (called the automated pressure evaluation system (APES) developed at IGER. The system allowed for 50 gas-tight culture bottles, each of which was fitted with a pressure sensor and solenoid valve linked to a computer, for continuous monitoring of all bottles. During fermentation, each solenoid valve opens to release the accumulated gas when the pressure sensor registers a pre-set gas pressure. The number of vents and time between each vent for each bottle are recorded automatically by a computer and are then used to plot cumulative gas profiles. This system is now also in place at Wageningen Agricultural University, and was used to produce the profiles for wheat straw and ryegrass shown in Fig. 10.6. The large number of data points produce very detailed fermentation profiles. This apparatus uses a combination of pressure transducer and an electric valve for every bottle. Each valve opening represented a fixed amount of gas and the fermentation kinetics were determined by recording the time of each valve opening on a data logger.

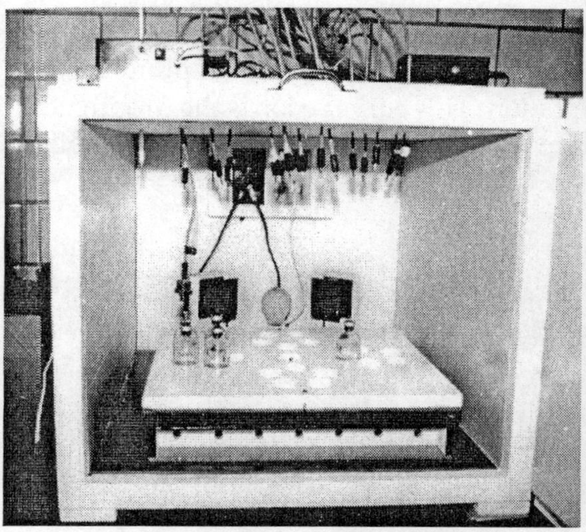

Fig. 3 : Equipment used for the automated gas system at Cornell University (photo courtesy of Alice Pell, Cornell University).

Fig. 4 : The IGER automated pressure evaluation system at Wageningen University, The Netherlands

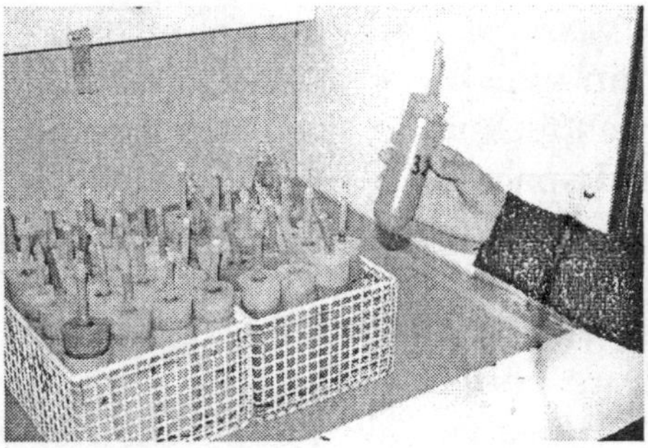

Fig. 5 : Digestion tubes containing samples for the rumen fluid/pepsin *in vitro* digestibility technique (Tilley and Terry, 1963).

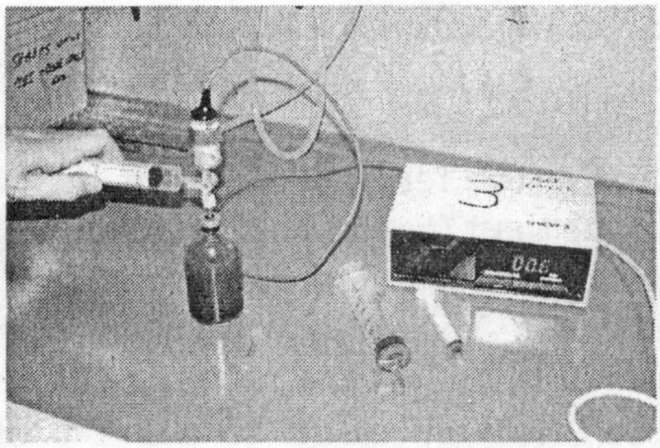

Fig. 6 : Gas production equipment. (After Theodorou and France, 1993.)

A general description of the method

Broadly speaking, the general principles of what the different techniques measure are the same. The technique is similar to other *in vitro* digestibility procedures which use milled substrates, an anaerobic medium and an inoculum of a mixed microbial population from the rumen. The pre-weighed substrate is suspended in a medium, the mixture warmed to ~39°C and a freshly collected sample of rumen fluid added as inoculum. From that moment, the production of gas resulting from fermentation is recorded, either at the end of the fermentation or at regular time intervals for those methods that measure fermentation kinetics. A cumulative gas profile is plotted from the results at the end of fermentation. This curve records the production of a waste product by the microorganisms, and so reflects the kinetics of microbial activity within the bottle.

Sample preparation

The preliminary screening of feed and fodder samples can be made by *in vitro* gas production technique. The fermentation may be conducted in the glass syringes containing the substrate and buffered rumen fluid at various times of incubation. The rate and extent of gas production can be estimated by fitting modified Mitscherlich equation in model of single exponential equation with and without lag phase to study the gas production kinetics of substrate.

As with all laboratory techniques, the preparation of a forage sample is fraught with many factors to be considered, which have been dealt with exhaustively for other *in vitro* methods. It is generally recognized that a smaller particle size increases the surface area available for microbial degradation. Consequently, for routine gas-production analysis most authors mill the test substrates through a 1 mm screen, in keeping with other *in vitro*. Most groups either freeze-dry samples (which, in the case of silages, can lead to the loss of volatile components) or oven-dry at a low temperature, either 60° or 70°C. Cone *et al.* (1994) reported a comparison of different drying conditions on the fermentation kinetics of grass and maize stems, and concluded that freeze-drying was better than oven-drying (at 30, 50, 70 and 105°C).

Substrate preparation

The feed substrates need to be evaluated are dried in a forced draft oven at 50–55 °C for 48 h, ground to pass a 1 mm screen and stored in air-tight polypropylene containers.

In vitro ruminal incubation

- The substrate (feed) is weighed on a plastic boat with removable stem and is placed into the bottom of the glass syringe without sticking to the sides of the syringe.
- The piston is lubricated with petroleum jelly and pushed inside the glass syringe.
- The syringes are kept in an incubator at 39°C until incubation.
- The following table gives details of solutions and the order in which they were added prior to the filling in syringes.

Preparation of buffer media

The composition of different solutions is given below.

1. Micro mineral solution

$CaCl_2 . 2\ H_2O$	13.2 g
$MnCl_2 . 4\ H_2O$	10.0 g
$CoCl_2 . 6\ H_2O$	1.0 g
$FeCl_3 . 6\ H_2O$	8.0 g

Dissolved in 100 ml of water

2. Buffer solution

NH_4HCO_3	4.0 g
$NaHCO_3$	35.0 g

Dissolved in 1000 ml of water

3. Macro mineral solution

Na_2HPO_4 anhydrous	5.70 g
KH_2PO_4 anhydrous	6.20 g
$MgSO_4$. 7 H2O	0.60 g

Dissolved in 1000 ml of water

4. Resazurine solution

0.10 % w/v

5. Reduction solution

1N NaOH	2.0 ml
Na_2S. 9 H_2O	285 mg
Water	47.5 ml *aqua dest*

All the solutions may be prepared well before the start of incubation except the reduction solution, which should be prepared freshly each time shortly before the rumen fluid was collected. The solutions will pour into a Woulff flask, mixed with a magnetic stirrer and warmed to 39°C in a PVC water bath with digital thermostat in the following order:

Solutions	Quantity (ml)
Distilled water	365
Micro mineral solution	0.1
Buffer solution	183
Macro mineral solution	183
Reduction solution	38.8
Rumen liquor	330
Total volume	**1100**

CO_2 gas should be passed through the submerged tube in the Woulff flask continuously during buffer media preparation. On the day of incubation, the mixture of rumen liquor and particulate matter (*approximately 60:40*) should be collected from the fistulated donor animal (before feeding) into pre-warmed CO_2 filled thermos and carried to the laboratory. The rumen fluid must be bubbled with CO_2 gas for few minutes and then add it to the buffer media when it will become colorless. All handling of rumen liquor should be done under continuous flushing with CO_2.

For ease in preparation of different set of glass syringes in the desired plan of experiments, the following table can be followed.

Items	30 syringes	45 syringes	60 syringes
Solution:I			
Distilled water	365 ml	550 ml	730 ml
Micromineral solution[a]	0.1 ml	0.15 ml	0.185 ml
Rumen buffer solution[b]	183 ml	275 ml	365 ml
Macromineral solution[c]	183 ml	275 ml	365 ml
Resazurin solution[d]	0.95 ml	1.45 ml	1.90 ml
Solution:II[e]			
1N NaOH	1.6 ml	2.4 ml	3.1 ml
$Na_2S.7H_2O$	220 mg	330 mg	440 mg
Distilled water	37 ml	55 ml	73 ml
Solution:III			
Rumen liquor	330 ml	500 ml	660 ml

Filling of syringes and incubation

- Before adding rumen liquor the colour of the medium had changed from blue over pink to colourless, indicating complete reduction of resazurine.
- Once the medium became colourless, the required amount of filtered rumen liquor is added. The proportion of medium mixture solution to rumen liquor is 2:1 for 30 ml and 3:1 for 40 ml incubations which is pumped with an automatic pipette into each syringe, prewarmed at 39^0 C.
- The syringes are shaken gently and residual air or air bubbles if any is removed and the outlet is closed.
- The level of piston is recorded and the syringes are placed in a waterbath ($39^o \pm 0.5^oC$).

- The syringes are shaken every 30 minutes for first 2 h from the start of the incubation and thereafter every 2 h up to 10 h of incubation.
- The blank values for gas production (with or without treatments) at each hour of incubation is subtracted from the volume of gas produced from the sample at that particular hour to get the actual gas produced.
- For estimating the rate and extent of gas production, the volume of gas produced is usually recorded at 0, 1, 3, 6, 12, 24, 30, 36, 48, 60, 72 and 96 h. Automatic recorders are also available for more frequently recording of gas volume at different incubation hours.

The fermentation kinetics can be estimated by means of nonlinear regression as follows.

(a) $Y = b \times (1-e^{-ct})$

(b) $Y = b \times (1-e^{-c} \times {}^{(t-L)})$

Where

y = gas production (ml) at time t

b = potential gas production (ml)

c = fractional rate of fermentation

L = discrete lag time (h)

The average fermentation rate (ml gas h^{-1}) may be defined as the average gas production rate between the start of incubation and the time at which the cumulative gas production was half of its asymptotic value, and may be calculated

Rate = b (c/2 (In2+cxL))

Substrate specific times are defined by the half time ($T_{1/2}$) of asymptotic gas production and calculated as

$T_{1/2} = \ln 2/c$

Voluntary feed intake potential (g DM kg^{-1} $BW^{0.75}$) can be calculated from *in vitro* gas production kinetics parameters using the equations

Intake (g DM kg^{-1} $BW^{0.75}$) = 18.90 + 0.23 (a+b) +687xc+0.11x CP

3.1.4. In vitro ruminal nitrogen degradation (IVDN)

The method of estimating ruminal protein degradation *in vitro* is based on measurement of ammonia nitrogen concentration and gas production. The method uses the known relationship between fermentation of carbohydrates and microbial protein synthesis for determination of NH_3-N incorporated in to the microbial proteins. The total amount of NH_3-N liberated is found by extrapolation of linear regression to zero protein synthesis.

- Incubations were carried out in 100 ml calibrated syringes according to the method described by Menke *et al.* (1979).
- Sample weight was chosen so as to correspond 26 mg crude protein. Hundred mg starch (an energy supplement) was incubated alongwith as an aid to microbial protein synthesis.
- For each sample 21 syringes (3 each for blank, blank + treatments, sample, sample + starch, sample + treatments, sample + starch + treatments and standard) were incubated. Rest of the procedure was essentially the same as described earlier.
- The incubation was terminated at 24 h and when gas production exceeded 80 ml, the piston was pushed back to the initial position.
- At the end of incubation (24 h) the amount of gas produced was measured by reading the position of piston.
- The contents of the syringes were transferred in to a 100 ml beaker on ice.
- A 10 ml of aliquot was transferred to the Kjeldahl distillation tube for determination of NH_3-N by distillation with 2 ml of 1N NaOH. The liberated NH_3 was trapped in 25-30 ml of boric acid and titrated with N/100 H_2SO_4 as per micro-kjeldahl distillation method.

Nitrogen degradability was calculated from the linear regressions of NH_3-N concentration (y; mg) vs. gas production (x; ml) observed on incubation of feed with and without endogenous energy source (starch). The intercept at Y axis (bo) represents the amount of NH_3-N which is released when no fermentable carbohydrate is added and hence no synthesis of bacterial protein takes place. The difference between the Y intercept (bo) and NH_3-N content in the blank indicated the amount of NH_3-N liberated from protein and other nitrogen containing compounds of the feedstuff incubated. The *IVDN* at 24 h was calculated as a proportion of total nitrogen incubated by the equation:

$$\text{IVDN} = \frac{\text{NH}_3\text{ - N at zero gas production }\left(b_0\right) - \text{NH}_3\text{ - N of blank}}{\text{Total nitrogen of feedstuff incubated}}$$

3.1.5. True digestibility

True digestibility was estimated as per the method outlined by Goering and Van Soest (1970) with little modifications.

- The incubations were carried out in 100 ml graduated glass syringes.
- The sample weight was increased to 400-500 mg to increase the residue so as to reduce the analytical error inherent to gravimetric determinations.
- To maintain pH and buffering for higher short chain fatty acid production, double strength buffer was used. The increase in buffer volume was matched by an equal increase in the volume of distilled water so as to keep the osmolality similar to the one in routine incubations.
- Forty ml buffered rumen liquor was injected in to each syringe.
- After 24 h incubation, the contents of the syringe were directly transferred in a 500 ml spoutless beaker. The syringe was washed with 50 ml of double strength NDS and washings added to the beaker.
- The contents in the beaker were refluxed for 1 h, filtered under vacuum through preweighed sintered (G1) crucibles. NDF content of the residue was determined.
- True digestibility was calculated as follows.

$$\text{TD} = \frac{\text{wt. of sample - wt. of residual NDF}}{\text{wt. of sample}} \times 100$$

- True digestibility of DM and OM was determined from the DM and OM content of sample and residual NDF. Partition factor was calculated as the ratio of substrate truly degraded to gas volume produced.

Do's and don'ts for the gas method

- The plunger of the glass syringe should be properly lubricated using white Vaseline (apply less amount of Vaseline for incubations up to 24 h and more for incubations up to 96 h).
- Collect rumen liquor from both the liquid and the solid phase and handle it properly (use of warm containers, flushing the containers with carbon dioxide, always keeping the rumen liquor under carbon dioxide).
- Reducing solution should be prepared fresh on the same day of conducting the experiment.
- Start flushing the medium with carbon dioxide well before (approx. 10 min) adding the reducing solution. Also flush the medium for at least 10 min after adding the rumen liquor and before starting filling the syringes. Keep flushing the medium with carbon dioxide while filling the syringes (the flow could be reduced at this stage).
- While filling the syringes with the medium, keep an eye on the medium (carbon dioxide gas should be flushing into the medium and the medium should be stirring).
- After dispensing 30 ml of the medium into the syringe, create a light vacuum by pushing back the plunger and then open the clip. This procedure will bring the medium lying in the nozzle back into the syringe. Otherwise there could be a loss of the medium and/or sample.
- After filling of the syringes has been completed (might take 30-40 min), shake the syringes. Shake them again after every 30 min till first 2 h of the incubation, and then after every two hours till the first 10 or 12 h of the incubation. Thereafter, shake the syringes after taking the gas volume readings (24, 30, 36, 48, 60, 72, 96 h, as the case may be). Make sure that all feed particles are taken into the medium while stirring (swirling shaking action might help).
- Wash the dispenser with distilled water immediately after finishing filling the syringes, otherwise the dispenser could get stuck up and might not then be usable.
- Check temperature and level of water in the water bath at least twice a day.

- In the evening before going home, if the plunger is above 80 ml level, push it back, record the readings (before and after) pushing back the plunger.
- When you push back the syringe in the evening, give a shake after approx. 30 min in order to prevent taking up the sample along with the bottom portion of the plunger and out from the incubation medium.
- Use the carbon dioxide gas cylinder with caution. Ask someone if you do not know its operation. Misuse could cause an accident.
- Keep the syringe in inverted position and in parallel with eye while recording the gas reading. Immediately transfer the syringe into the water bath after taking the reading.
- For cleaning the syringes, the syringe should be emptied (preferably pulling back the plunger and removing contents from the back and not from the nozzle). Clips should be removed. The plunger and the outer graduated part of the syringe (barrel) should be separated. Excess Vaseline on the plunger should be cleared with a tissue paper or a piece of soft cloth, and then transfer both the parts in hot water containing detergent (soap) solution. Rub the plunger with hand and inside of the barrel with a soft brush to clean these. Wash thoroughly both the portions with hot water and finally rinse them with distilled water. Dry them well before weighing sample into the syringe.
- Fix the clip in such a manner (by keeping the portion, where pressure is applied to open or close it, facing the syringe) that it does not open by striking on the edges of the lid of the water bath while taking out the syringe for taking reading.
- Mark the crucibles well (preferably with a diamond pencil). Keep them in increasing or decreasing order; this might help you in identifying the crucibles, which have not been marked well, especially after these have been placed in the Muffle furnace.

3.2. Use and Significance of the Technique in Feed Evaluation

Media

The living rumen provides an ideal environment for its anaerobic microorganisms in terms of pH, temperature, buffering capacity and nitrogen in the form of urea, while the microorganisms behave symbiotically by providing each other with micronutrients, such as branched-chain fatty acids and vitamins. There are also likely to be other unknown factors present in the rumen milieu, which cannot yet be replicated *in vitro*. In the past, quite a number of media used to characterize rumen microorganisms overcame this problem by addition of a large proportion of sterile rumen fluid to ensure the presence of all possible factors for microbial growth. All the media in current use have the following in common: bicarbonate and phosphate buffers, a reducing agent, a source of nitrogen, various minerals and resazurin as an indicator of redox potential. In all cases, CO2 is used during the preparation of the medium to ensure a low redox potential at the time of inoculation. A low redox potential is very important for the fermentation of very fibrous forages, or even for those rich in starch. The absence of strict anaerobiosis resulted in partial losses of cellulolytic and amylolytic bacterial groups from rumen fluid. Therefore a low redox potential is very important, not only for the fermentation of fibrous forages, but also for those rich in starch. It is recommended that all *in vitro* systems used to assess fibre digestion kinetics must use continuous CO2 gassing and reducing agents to promote both the rapid initiation and the progression of fermentation. Minimizing the limitations of the *in vitro* system means that differences between substrates will be more readily detected. A modified version of medium B is used most commonly in Wageningen; it is a complex semi-defined medium, which, in principle, should provide most of the rumen microorganisms with all of their requirements, except for energy, which is provided by the test substrate. This medium, while suitable for a wide range of the rumen microorganisms, comprises several component solutions and so is time-consuming and complicated to use for routine analyses. Not surprisingly, several other media have been developed, most of which seem to originate either from the medium of germen workers or from that of english lab. Research continues to develop media that are appropriate for different situations.

The composition and strength of the buffer are still a focus of interest. Measurement of pH at the end of fermentation is to be recommended to ensure that sufficient buffering capacity was available for novel or concentrate substrates.

Inoculum

Conditions for the keeping of donor animals, their feeding, and the timing of rumen fluid collection, so as to obtain the most consistent result between runs for *in vitro* techniques was reviewed for *in vitro* techniques by Alexander and McGowan (1966). It was recommended that a sample should be taken after overnight fasting from at least three animals, all of which were fed the same diet of medium-quality hay with a small amount of concentrate.

Animal species

Sheep are most often used as a routine rumen fluid donor for *in vitro* techniques. They are chosen mostly for their size, which makes them cheaper to feed and house for an extended period. However, cattle are also used as inoculum donor. Differences in the activity of microorganisms from rumens of different species or of the same species but different diets mean that all descriptions of *in vitro* gas-production evaluations should describe the conditions of the donor animal (species and breed, diet, time of collection in relation to feeding, etc.).

Time of collection

Rumen fluid taken after overnight fasting is less active than that taken 2 h after feeding, but is more consistent in its composition and activity. As a rule, therefore, it has been recommended that rumen fluid be collected before feeding, which is the routine followed by many groups.

Preparation of inoculum

Another important consideration is that of homogenization of the rumen fluid before inoculation. It has been shown that blending increases the number of adherent bacteria (which are most often the cellulolytic species) in the inoculum, but it can also increase the number of small feed particles present, which can lead to higher gas production in the blank bottles. The blending procedure can also increase the risk of oxygen exposure if the stream of CO2 is not forceful enough.

Proportion of inoculum to medium

The proportion of inoculum to medium has been investigated and it is recommended an inoculum of 20% of a total volume of 10 ml. However, the amount of added inoculum varies widely between groups. In response to increased bacterial density rates of gas production from forages was high.

What does the Technique Measure?

In vitro gas production is related solely to fermentation in the rumen, rather than to digestibility in the whole digestive tract, which also includes enzymatic digestion, absorption and hind-gut fermentation. Rumen fermentation is mainly the result of the activity of the bacteria, protozoa and fungi that inhabit it. Some species from each of these groups possess cellulolytic and hemicellulolytic activity, though the relative contribution of each to fermentation is still not completely understood.

3.3. Stochiometry of Rumen Fermentation

Energy for the growth of rumen microorganisms is derived from the fermentation of carbohydrates, particularly starch and cell walls, which are first broken down into hexoses and pentoses, from which the main end-products are acetate, propionate and butyrate (VFA), CO2 and CH4 , but which can also supply essential carbon skeletons for the synthesis of microbial biomass. The VFA are absorbed across the rumen mucosa and provide energy for the mucosal tissue, as well as for the host animal. The gases produced from VFA production and from the bicarbonate buffer in the rumen are eructated to the atmosphere. It is this gas that is being measured *in vitro*. The stoichiometric reactions of fermentation of cell-wall hexoses were described as below:

1 mol Hexose + $2H_2O$? 2 Acetate + $2CO_2$ + $4H_2$ (1)

1 mol Hexose + $2H_2$? 2 Propionate + $2H_2O$ (2)

1 mol Hexose ? 1 Butyrate + $2CO_2$ + $2H_2$ (3)

CO_2 + $4H_2$? CH_4 + $2H_2O$ (4)

Formation of propionic acid is the only reaction that requires the uptake of hydrogen (H2). Any remaining H2 is usually converted in the rumen to methane by methanogenic bacteria. Also, the production of propionate does not involve the generation of CO2. Therefore, to compare gas data between different forages, it is also useful to examine differences in the acetate;propionate ratio at the end of fermentation. The complete pathway of VFA production for the hexose equivalents from high-forage diets was summarized and given as:

1 mol Hexose = 1.34 mol acetate + 0.45 mol propionate + 0.11 mol butyrate + 0.61 mol methane + 4.62 mol adenosine triphosphate (ATP)

All techniques, except for the Hohenheim gas test, rely on the measurement of pressure in a receptacle of fixed volume. The empirical gas laws of Boyle and Gay-Lussac, from which the volume of gas can be calculated are:

$PV = nRT$ (5)

where:

P = pressure of gas (atm)

V = volume of gas (l)

n = number of moles of gas

R = the molar gas constant at 0°C and 760 mm Hg

T = temperature (°K)

In most studies, it is assumed that the pressure of gas is at 1 atm, so 1 mmol of gas occupies 25.62 ml at 39°C. However, if laboratories are located in areas where the atmospheric pressure is not 1 atm, an adjustment must be made accordingly.

An important aspect of gas production of the anaerobic system is that stoichiometric laws of fermentation balance can be applied since fermentation products must be derived from substrate incubated. High correlation between stoichiometrically calculated gas and actually recorded values have been reported. The stoichiometric balance allows the theoretical calculation of equilibrated amounts of the products (SCFA and gases). If the molar proportion and amount of SCFA are known, the theoretical amounts of CH_4 and CO_2 expected from the rumen fermentation can be calculated.

The need to evaluate the nutritive characteristics of ruminant feeds must increase in importance in coming years. Determination of the fermentation kinetics of substrates for their routine evaluation is likely to be an important facet of evaluation systems of the future, particularly as the dynamics of nutrient utilization are taken into account. Synchronization of energy and nitrogen supply to the rumen is considered to be an important approach to improve the efficiency of rumen fermentation. To apply this principle in ruminant feed formulation, one needs sufficient data on rates of protein degradation and carbohydrate fermentation by rumen microbes. It is evident that energy supplements differ in their rate and extent of ruminal fermentation and cumulative gas-production techniques could be used to generate such information *in vitro*.

Cumulative gas production is being used in combination with alkane analysis of forages as a measure of grazing intake in dairy cows. This methodology could be extended further to investigate the value of feed consumed by wild animals, by analysing the alkanes present in their faeces and then carrying out a fermentation study of the plants that are then known to have been eaten. In addition to routine screening and feed evaluation, the technique is also being used to study more fundamental aspects of gastrointestinal tract (GIT) fermentation, particularly in relation to plant cell walls and other fermentable carbohydrates. Valuable information is being obtained, particularly when used in combination with other techniques already available. Recently cumulative gas production in conjunction with cell-wall sugar analysis at various time intervals was used to evaluate differences in the fermentation of different anatomical components of soybean cell walls. Given the flexibility of the technique in terms of medium chosen and inoculums used, it is also being adapted to investigate more fundamental questions relating the source of substrate to the amount of microbial biomass produced. For this kind of research, it is essential to remember that one is not only examining the test substrate, but also microbial behaviour in terms of that test substrate and the medium being used. For example, when investigating the relation between biomass and gas production, it must be realized that the microbes must have access to all possible micronutrients they may require, or the results from an *in vitro* test will bear little resemblance to what occurs *in vivo*. Progress will continue to be made in terms of the technical details of the methodology. However, it is clear that once these details have been refined, the technique can be used to measure the dynamic interaction that exists between the substrate and microbes for fermentation in the rumen or, indeed, the rest of the gastro-intestinal tract. Results of fermentation kinetics are likely to change our perceptions of both feed evaluation and the kinetics of microbial metabolism.

3.4. Microbial Biomass Production

It has been shown that an inverse relationship can exist between gas volume (or VFA) and microbial biomass production. The type of carbohydrate and its rate of fermentation could influence total synthesis of microbial protein per unit volume of gas produced. On the one hand, it has been suggested that the efficiency of microbial growth can vary depending on feeding conditions. Yet it has also been pointed out that, if there are no deficiencies of microbial growth factors, both the source of carbohydrate (whether sugar, starch or fibre) and the flow rate of liquor from the rumen have little effect on microbial growth yield. The methodology used to measure cumulative gas production can easily be adapted to investigate the effect of microbial growth factors and sources of carbohydrates under controlled conditions, which could help to answer some of these questions. Little work has been done to examine the relation between gas production and microbial growth using the Hohenheim system. it was found that some substrates with proportionally high gas volumes had comparatively low biomass yields. Gas volume measurements were complemented by a determination of the truly degraded substrate *in vitro* to avoid selection of substrates with proportionally high VFA production and low microbial biomass yield, which is relevant for forages with a low nitrogen content. It has been pointed out that one of the factors affecting microbial growth efficiency is the relative amounts of carbohydrate and protein that are fermented. For example, a high protein-to-carbohydrate ratio could lead to a relatively low microbial protein-to- VFA ratio. These relationships could be examined more closely using cumulative gas-production techniques, particularly if other parameters, such as VFA, ammonia (NH3) and microbial protein production are also measured, using a medium that is known to contain all possible microbial growth factors, or, contrarily, is deficient in one or more of them.

3.4.1. Determination of microbial mass

Current nutritional concepts aim at high microbial efficiency, which cannot be achieved by measurement of gas only. The relationship between SCFA and microbial mass production is not constant and the explanation for this resides in the variation of biomass production per unit ATP generated. This can impose an inverse relationship between gas volume (or SCFA production) and microbial mass production particularly when both are expressed per unit of substrate truly degraded. This implies that selecting roughages by measuring only gas using *in vitro* gas methods might result in a selection against the maximum microbial mass yield. Blümmel *et al.* (1997) have demonstrated how a combination of *in vitro* gas production measurements with a concomitant quantification of the truly degraded substrate provides important information about partitioning of fermentation products, and the *in vitro* microbial mass production can be calculated as:

Microbial mass (mg) = mg substrate truly degraded - (ml gas volume × stoichiometrical factor)

For roughages, the stoichiometrical factor was 2.20 and the range is 2.20 to 2.34 mg/ml for high acetate and high propionate producing feed types.

3.4.2. Partitioning factor

The parameters in the above equation also allow the calculation of a partitioning factor (PF), which is defined as the ratio of substrate truly degraded *in vitro* (mg) to the volume of gas (ml) produced by it, i.e.

Microbial mass (units) = gas volume (PF - stoichiometrical factor)

A feed with higher PF means that proportionally more of the degraded matter is incorporated into microbial mass, i.e., the efficiency of microbial protein synthesis is higher. Roughages with higher PF have been shown to have higher dry matter intake. Thus, the PF calculated *in vitro* provides meaningful information for predicting the dry matter intake, the microbial mass production in the rumen, and the methane emission of the whole ruminant animal. It seems therefore justified to suggest that feeds or feed ingredients should be selected that have a high *in vitro* true degradability but low gas production per unit of truly degraded substrate. The theoretical range for PF is 2.74 to 4.41. Any value above or below this range should be critically evaluated and for tannin-rich samples, the PF is normally above 4.41.

3.4.3. Partitioning factor (PF) and Metabolisable energy

PF may be calculated as the ratio of mg of TDOM to ml of gas thereby produced. Metabolisable energy (ME, MJ kg^{-1} DM) may be estimated from the gas volume using the equation

ME (MJ kg^{-1} DM) = 2.20+0.136´gas (ml 200 mg^{-1} DM)+0.0057´CP (%DM)

3.4.4. Some useful tips in assessing rumen microbial protein synthesis

Many ways of arriving at microbial protein synthesis

1. Microbial mass (DM) = ADDM - TDDM

   ADDM (Apparently Degradale DM) = Undegraded residue - Blank

TDDM (Truly Degradable DM) = Undegraded residue after NDS Fluxing

2. Microbial N = TN – (NDF-N + NH_3-N)

   TN = Total N = Feed N + N in buffered RL before incubation (termed as blank at 0 h)

   NDF-N = N in undegraded NDF residue

   NH_3-N = NH_3-N in the superntant after 2 hr incubation

   N:B: Amino acids & peptides remain after incubation is considered negligible.

3. Microbial N = APUR-N - NDF-N

   APUR-N = Apparent undegraded residual N after incubation corrected for blank (0 h)

   N.B.: MN =(Feed N ± Change in NH_3-N) – NDFN

   Change in NH_3-N = NH_3-N in supernatant (BRL) at 0 h – NH_3-N after incubation

   If NH_3-N at oh > NH_3-N post incubation the change is added into FN & when NH_3-N 0hr < NH_3-N post incubation the change is substracted from FN

4. Microbial purines in ADDM

5. Purine derivatives in Urine

Useful tips

(i) N Content of rumen microbes 8.5% (Schafft, H. 1983, Untersuchung zur Quantifizierung des Abbans von Futterprotein in den Vormagen von Milchkuhen, Diss, University of Gottingen, Germany)

(ii) N Content in Microbial DM 7.7% (Orscov, E.R. and Ryle , M. 1990. Energy Nutrition in Ruminants. London, Elsevier)

(iii) Microbial Mass = Purines equivalent to RNA x 5 (20% RNA/ DNA in rumen microbes)

(iv) Microbial Cell = Polysaccharide(16.6) + protein (52.4)+ lipid (9.4)+ RNA(15.7)+DNA (3.2)

(v) $$MN(g/day) = \frac{\text{Purine derivatives absorbed}(mmol/d) \times 70}{0.116 \times 0.83 \times 1000}$$

= 0.727× PD(mmol/d) absorbed

0.83 = Digestibility of microbial purine

70 = N content of purine (mg N/mmol)

0.116 = Rate of purine-N to total N in mixed rumen microbes (11.6/100)

(vi) Purine derivatives excreted/d = PD absorbed/d×0.84+(0.150×LW $^{0.75}$ e $^{-0.25}$ ×PD absorbed/d)

(vi) RNA contains 22% purines, DNA contains 20.5 % purines

(vii) Avg. N content of purines is 49%

(viii) RNA contains 10.8 % N

(ix) Bacterial protein contains 15 % N = N x 6.67

(x) *Adenine* : Guanine ratio for yeast RNA = 1:1.3

(xi) *Purine* : Protein ratio of pure culture of predominant rumen bacteria = 0.0883

Purine : Protein ratio of mixed bacteria samples isolated from ruminal contents = 0.0306

(xii) Mixed ruminal bacteria has an average of 51.4% protein (range 32.2- 70.6%)

(xiii) Mixed ruminal bacteria isolated from contents will be having 7.15 % RNA which equals to 20.63% in pure rumen bacteria culture

(xiv) Protozoa do not contribute to post-ruminal protein supply in proportion to their contributions to the total microbial mass in the rumen.

(xv) Microbial CP as provided by bacteria and protozoa is considered to contain 80% true protein & 20% nucleic acids.

3.5. Data Processing and Interpretation of Results

Conversion of raw data to cumulative volumes

Gas volumes recorded at each time interval are added together to give a cumulative gas profile. In the case of the measurement of head-space pressure and volume, a correction is made for irregularities in head-space volume between bottles, by a regression of pressure against volume for each bottle. A correction is also made to cumulative volumes for the starting DM/organic matter (OM) weight of the substrate. Ideally, if remaining substrates are not required for further analysis, OM data should be used, since ash makes no significant contribution to gas or VFA production and can vary significantly between substrates. The yield of gas or VFA (*Y*G or *Y*VFA: the amount of gas or VFA produced per gram material disappeared) can also be calculated (as DM or OM) if the substrate losses at the end of fermentation are measured.

Correction for blanks

A series of blank bottles containing medium and inoculum but no substrate is routinely included in each run. The total gas recorded for the blanks is subtracted from the total gas produced from the test substrates to obtain the total gas derived from fermentation of the substrate. However, the blank profile is not subtracted from the substrate profiles prior to curve fitting, particularly in the case of the multiphasic curves, as the phases present in the blank may not be relevant to what occurs in the presence of substrate.

Curve fitting

Fitting cumulative gas profiles to an appropriate equation allows one to summarize the kinetic information therein. Several equations have been developed for this purpose. The choice of equation to fit gas profiles must take into account not only mathematical considerations (such as the statistical fit) but also the biological meaning of the parameters. Increasingly frequent readings for gas production lead to very detailed profiles, and considerations of mathematical fit (particularly very low Durbin–Watson values) have led to the development of a multiphasic model, which was based on the multiphasic growth model. The potential value of this model is that it separates the test substrate into the more rapidly and more slowly fermentable fractions.

Comparison have been made using different equations available in mathematical terms, and also with same equations, using a large series of data sets from widely different forages, and have analyzed the results from the point of view of the use of the equations in practice.

3.6. Applications of the Technique

Advantages and Limitations of the Technique

Cumulative gas-production techniques have several major advantages in terms of animal welfare, sample size and cost and, most importantly, in the fact that they describe the kinetics of microbial activity in response to a given substrate.

Animal welfare

While the use of animals is far less than is required for *in vivo* digestibility trials, gas-production techniques still require animals as a source of rumen fluid. To eliminate this need for fistulated animals, some investigations have been carried out to determine whether or not faeces could be used as an alternative inoculum. A comparison between rumen fluid and faeces from cows fed a silage diet showed that the faecal inoculum resulted in significantly lower total gas, DM loss and VFA production for all substrates (wheat straw, rye grass, soybean hulls, wheat, maize and soybean meal). Consequently, it is not yet clear whether faeces would be an ideal replacement for rumen fluid for this method.

Sample size

The amount of material required to assess the fermentation kinetics of substrates ranges from 0.1 to 1 g. More is required than is actually fermented, in order to ensure a representative sample, but these are still small amounts compared with other techniques, making it suitable for materials available in limited quantities (e.g. from plant-breeding trials, plants grown under special conditions, etc.).

Costs

The costs of laboratory tests are usually much lower than those required to maintain animals. Whether the manual or automated systems are cheaper to run depends, to a large extent, on the cost of labour in the country where the laboratory is located. The initial equipment required can be quite expensive for the automated systems, but, as the hours for measuring are significantly reduced, it is a matter of balancing the capital costs of the equipment with the labour costs.

Information obtained

Cumulative gas production measures the kinetics of fermentation (i.e. microbial activity), rather than a reduction in particle size.

Reduction of particle size is an indication of substrate degradation, but does not necessarily indicate the availability of material to microorganisms. However, gas production remains an indicator of what occurs in the rumen, rather than in the whole animal.

Limitations

The limitations of the gas-production technique rest, for the most part, with the expectations of those using the technique. If someone hopes that the technique can be used to predict all aspects of animal health and behaviour, he/she will be disappointed. However, it does measure the kinetics of fermentation of a given substrate with a given microbial population, thereby giving a reasonable simulation of what occurs in the rumen, and, as such, will be a useful tool for evaluation of ruminant feedstuffs. One limitation to the use of the technique for forage evaluation is the lack of uniformity in methodology. This makes it difficult to compare results from different groups. An urgent task facing those who are using the technique is to remedy this situation. It should be possible to standardize the technique in terms of the kind of medium being used, the timing of inoculum collection and the diet of the donor animals. Steps have been made in this direction. Two ring tests were organized in 1994 and 1996 (Agricultural Development and Advisory Service (ADAS) – Jim Huntington and Caroline Rymer, personal communication). Preliminary results from the first test, in which all groups used their own equipment and methodology, were reported at an informal meeting held at the University of Reading in 1995, while results of the second test, in which the medium, inoculum collection and, to some extent, animal diet were standardized, have also been reported. The use of standard substrates, as pioneered for the Hohenheim gas test, may need to be considered in order to overcome the inevitable differences between inocula, not only between groups, but also between runs. Other important details which still need to be investigated include: medium (e.g. an ideal buffer concentration, presence of micronutrients and anaerobiosis) and treatment of inoculum to maximize cellulolytic organism activity; the use of blanks; sample preparation in terms of method and duration of drying and grinding; and, very importantly, the relationship between gas, VFA and m i c r o b i *in vitro* system. Inevitably, compromises will have to be made between what would be ideal in terms of the metabolism and growth requirements of microorganisms and what is practical on a day-to-day basis. Some flexibility within the system will need to be maintained to allow for the requirements of different research questions, such as the potential ATP and/or biomass yield of forages.

3.6.1. Some practical considerations for feed evaluation using the gas method

- The termination of the incubation should be at the time at which the efficiency of microbial production is maximum, which is difficult to ascertain under normal feed evaluation experiments for a large number of feeds. An approach that could be applied is running of 200 mg sample for the kinetic experiment and calculating the time (t/2) at which the gas production is half of the potential gas production. The time 't/2' is near to the time at which the efficiency of microbial production is the maximum. This means that one has to run a kinetic experiment before determination of truly degraded organic matter and microbial protein using purines. The time 16 h for concentrate feeds and 24 h for roughages might be a good compromise for routine feed evaluation

- The method of measurement of undegraded residue does not work satisfactorily with tannin-containing feeds/samples (presence of tannin-protein complexes as artifacts in the residue) and starch-rich feeds (some starch might not be degraded by microbes up to 16 h of incubation but it would get solubilized in the neutral detergent solution leading to underestimation of truly undegraded residue).

- It may be noted that this method of measurement of undegraded residue should not be applied at the initial hours of incubation (before 16 h of incubation) since during this period a portion of the feed, which is undegraded by microbes could be solubilised in the neutral detergent solution.

- The high PF values for tannin-rich feeds appear to be due to the solubilization of tannins from the feeds during fermentation, contributing to the dry matter loss but without contributing to the gas, and inhibition of gas production from cell solubles by tannins through inhibition of rumen fermentation (cell solubles will also contribute to dry matter loss). The presence of tannin-protein complexes in the truly undegraded residue and hence underestimation of the truly degraded organic matter also contribute to the errors in the PF, but the presence of these complexes in the truly undegraded will lower the PF. The PF, which is obtained for the tannin-containing sample is the net effect of solubilization of the feed components during fermentation but not contributing to the gas production and the presence of tannin-protein complexes in the truly undegraded residue.

- Estimation of in vivo metabolizable energy (ME) and digestibility of organic matter (DOM) using the net gas value from 400-500 mg sample and converting the gas value from to 200 mg dry sample or directly incubating 200 mg sample for 24 h.
- Microbial N (MN) could also be measured after incubation by following two nitrogen balance approaches (Getachew et al., 2000). The first approach is:

  MN = TN – (NDF-N + Ammonia-N),

  where TN is total N i.e., feed N + N in buffered rumen fluid in the syringe before incubation (at 0 time),

  NDF-N is the N bound to NDF fraction following incubation and

  Ammonia-N is the ammonia-N in the supernatant following the incubation.

In a closed system, the total N present at the start of the incubation can be in microbial mass, NDF-N, ammonia-N and amino acids during any time of the incubation. Negligible amounts of amino acids and peptides are present in the supernatant during fermentation and therefore these can be ignored in calculation of microbial N.

For determination of NDF fraction for N bound to NDF after incubation, the syringe contents are transferred into a 600 ml beaker and the syringes are washed twice with a total of 50 ml neutral detergent solution (NDS) and emptied into the beaker. The contents are refluxed for 1 h, and then filtered through pre-tarred filter crucibles (no. 2). The crucibles are dried overnight at 100° C and weighed. True degradability can be calculated as the weight of substrate incubated minus the weight of the residue after NDS treatment. The residue after NDS treatment (neutral detergent residue, NDF) is subjected to micro-Kjeldahl digestion for determination of NDF-N.

The second approach is: MN = APUR-N – NDF-N,

where APUR-N is N bound to apparent undegraded residue after incubation.

The preparation of apparent undegraded residue is given above (section 'd'). N bound to this fraction is determined in a manner similar to NDF-N determination. From the weight of apparent undegraded residue, apparent digestibility can also be calculated.

Research and development efforts are required to establish a feed library for unconventional feedstuffs that includes information on nutritive values in addition to routine composition analysis. In the case of tannin-containing feedstuffs, there is a need to incorporate approach(s) measuring the biological activities of tannins as well as measuring tannin levels by chemical methods. Further studies are required on: i) the development of simple approaches for identifying the incubation time in the *in vitro* gas system at which the PF (a measure of the proportion of fermented substrate which leads to microbial mass production) is maximum, ii) the effect of nitrogen in the incubation medium on the PF, and iii) the *in vivo* significance of the PF so obtained.

3.7. Enzyme Techniques for Estimating Digestibility

The development and application of laboratory methods for predicting the nutritive quality of forages have been an active and successful area of research during the latter half of this century. An important development has been the introduction of biological methods, using a rumen microbial inoculum as a digestion medium in place of the chemical analysis for constituents such as fibre, cellulose and lignin. Although methods using microbial inocula have been widely and successfully used for a range of different forages, there are inherent problems in their use. Attention has, therefore, been focused more recently on the replacement of microbial inoculum methods by enzyme techniques which use crude cellulase preparations.

3.7.1. The need for alternatives to rumen fluid

Methods of predicting in *vivo* digestibility using rumen microbial inoculum have become well established, particularly the two-stage in *vitro* technique of Tilley and Terry (1963). It has been shown that their method gives results well related to in *vivo* digestibility for a range of forages and under a variety of growing conditions. There are, however, inherent problems in the use of microorganisms in rumen fluid for forage analysis.

1. The activity of the rumen microorganisms is lost or altered by freeze-drying or other methods of preservation. Fresh rumen fluid

must therefore be readily accessible, which entails having fistula ted sheep or cattle available as donor animals.

2. The donor animals need to be maintained on a standard feeding regime for long periods to minimize variability in inoculum potency. There is therefore a high and costly labour input involved in the feeding and care of animals.
3. There are analytical difficulties in using rumen fluid, e.g. the need to maintain anaerobic conditions, the high viscosity of digesta precluding filtration, unpleasant smells and the need to observe strict hygiene to avoid possible risk of infection from pathogens.
4. Increasingly, there is concern over the use of surgically modified animals in experimentation, and alternative methodologies need to be used where appropriate.

The main disadvantage of using rumen microbial inoculum in routine analytical procedures, however, is the variability that is found in the potency of the inoculum from week to week. The factors influencing the potency of inoculum have been widely studied, particularly in relation to the Tilley and Terry technique. Significant variation in the activity of inoculum has been shown to occur between different donor animals and for the same animal on different days. It is suggest that the consumption of water by sheep at varying intervals prior to collecting inoculum may explain much of the variation in the potency of inoculum; different methods of processing inoculum have been proposed in attempts to reduce variability. The need for proper calibration, using microbial inoculum methods and regard the inclusion of appropriate standard forages of known in vivo digestibility within each analytical batch to be essential and have been stressed greatly. The inclusion of standards allows correction of the results relative to standard values; in practice, it is often difficult to make a valid correction unless a large number of standard samples are included in each analytical batch.

The use of cellulolytic enzymes as alternatives to rumen fluid is clearly an attractive prospect if their use gives results comparable to those of inoculum methods. The use of enzymes eliminates the need for fistulated animals and anaerobic procedures, thereby simplifying analytical methodology and eliminating the variability in activity between analytical batches.

3.7.2. Characteristics and activities required for enzyme techniques

The chemical constituents of forages can be divided into those making up the structure of the plant (cell-wall constituents) and those contained within the cell (cell contents). This partition can be conveniently defined analytically as neutral- and acid-detergent procedures. The cell contents are essentially completely digestible in vivo and comprise soluble carbohydrates, starch (in some forages), protein, organic acids, lipids and soluble minerals. The cell-wall constituents (essentially cellulose, hemicellulose and lignin) vary in digestibility depending on their polymeric configuration, degree of crystallinity and degree of lignification. Lignin itself is largely indigestible under anaerobic conditions and its presence exerts a profound effect on the digestibility of the cell-wall fractions associated with it. The in vivo digestibility of the cellulose and hemicellulose therefore varies from completely digested when unlignified to completely undigested when highly lignified; the cuticle and surface waxes are also indigestible to a large extent.

The enzyme activities sought for analytical procedures to predict in vivo digestibility need to reflect the digestive processes in the ruminant. The analysis needs to remove the soluble constituents (cell contents) and to solubilize unlignified and even moderately lignified cell wall to a significant extent. The extent of solubilization does not necessarily need to be as complete as in vivo but needs to reflect the effects of lignification in a comparable and correlated way.

Enzyme Techniques

Soluble constituents are readily removed in the pretreatments with acid pepsin or neutral detergent, which form part of most of the proposed enzyme methods. The presence of starch in certain forage crops (maize, cereals) needs to be taken into account, since it is not removed by acid pepsin, neutral detergent or incubation with cell-wall-degrading enzymes. An additional treatment with amyloglucosidase or mild acid hydrolysis is therefore required in the analysis of forages containing starch.

The cell-wall-degrading enzyme used must be sufficiently active to achieve the second objective, i.e. the solubilization of a significant amount of the cell wall. Crude 'cellulase' preparations are readily available in liquid or solid form; all appear to hydrolyse both cellulose and hemicelluloses (Table 8.1). There is a wide disparity in the activity of cellulases from different fungal sources (Table 8.2) and it is essential to establish the activity of the cellulase by reference to plant cell-wall material or cellulose paper and not to soluble derivatives, such as carboxymethylcellulose. Crude cellulases from *Trichoderma* species have generally been found to be the most reliable sources of crude cellulase, but the activity of enzymes from different suppliers should always be checked before use.

Sources of cellulolytic enzymes

Fungi and other microorganisms producing enzymes able to digest plant fibre are very widely distributed in nature, although those able to degrade crystalline cellulose are less common. The microbial degradation of cellulose and its related polysaccharides in soil is a vital process in the maintenance of the carbon cycle, enabling the decay of vegetation and the recycling of nutrients. Microorganisms in the gut of herbivores enable these animals to survive on diets largely composed of plant fibre; vast areas of natural and improved grasslands are thereby utilized for the production of food.

There is, therefore, a wide choice of potential sources of cellulases for use in forage analysis. Ideally, the enzyme system used to predict in *vivo* digestibility would be derived from rumen microorganisms. Cell-free extracts from disrupted mixed rumen bacteria have been shown to rapidly hydrolyse plant hemicelluloses, these extracts, however, have little or no activity against purified crystalline cellulose, although they do attack the amorphous polymer to varying degrees. Attempts to extract active cellulase preparations from rumen.

Hours of incubation

Microorganisms for use in enzyme assays have been largely unsuccessful in comparison with aerobic fungi and gave poor yields. This is undoubtedly related to the anaerobic growth habit of microorganisms in the rumen, which means that the yield of extracellular hydrolases is poor in comparison with that of their aerobic counterparts. Other significant factors in enzyme production include the fact that rumen bacterial cellulases are not extracellular but are often produced in a large complex, the cellulosome, which is tightly associated with the bacterial cell or bound to the substrate that is being degraded. The hydrolytic enzymes produced by the anaerobic gut fungi are an exception, however, and culture filtrates of these organisms do have considerable extracellular hydrolase activity, probably because they have very high specific activities. Nevertheless, the ability to produce significant quantities of gut fungal hydrolases is limited by the fact that these organisms are obligate anaerobes and large-scale industrial fermentation systems do not exist for their exploitation.

In view of the difficulties in obtaining cellulolytic enzymes from rumen sources, attention has been directed towards the aerobic cellulolytic fungi. The cellulase and associated enzyme systems of these fungi are extracellular and are therefore readily recovered from the culture fluid in which the organism is grown. Culture solutions comprise a mixture of salts, nutrients and cellulose at the requisite pH. After a growth period of 7-10 days under aerobic conditions, the culture fluid is removed by filtration and freeze-dried or an acetone powder is prepared. Suitable conditions for the preparation of an active cellulase-hemicellulase from Trichoderma species.

Trichoderma cellulases have generally been found to be more active than those of other fungi, when measured using native cellulose or plant cell walls as substrates. Extracts from Trichoderma have therefore been extensively studied to determine the nature of the cellulase complex. It is evident that the complex comprises several enzymes necessary for attack of highly orientated forms of cellulose. This includes exocellulase and endocellulases, which attack amorphous regions of the cellulose polymer, producing shorter-chain oligomers, and a cellobiohydrolase, which hydrolyses cellobiose to glucose. All components apparently need to be present for the efficient solubilization of cotton cellulose with complicated synergistic effects existing between the different enzymes. It is clearly essential that cellulases used for forage analysis retain all the components of the cellulase complex. Attempts by suppliers to purify or concentrate crude cellulase extracts may be responsible for the poor activity noted in cellulases from some fungal sources.

3.8. The Development and Validation of Cellulase-Based Techniques

3.8.1. Relationship of solubility in cellulase to *in-vivo* digestibility

The earliest attempt to replace rumen inoculum as a digestion medium for the laboratory estimation of forage digestibility appears to be that of Donefer *et al.* (1963). These workers found the solubility of forage in cellulase, or cellulase with pepsin, to be no better related to *in vivo* digestibility than solubility in buffer solution without enzyme; solubility in cellulase was better related to voluntary intake than to digestibility. It should be noted that the cellulase used by Donefer *et al.* (1963) appears to have a low activity and that the concentration of cellulase used was much lower than that used by subsequent workers.

The first successful use of cellulase as a replacement for inoculum was reported by several workers , using a crude cellulase derived from a soil basidiomycete. Jarrige *et al.* found solubility in the cellulase to be highly correlated (r = 0.92) with *in vivo* digestibility for grass and legume forages. Cellulase solubility gave a better prediction of digestibility than the Tilley and Terry inoculum method and was highly reproducible, with only a third of the standard error of the inoculum method. There was no significant week-to-week variation using cellulase but variability using inoculum was highly significant between weeks. It was shown that the crude cellulase also contained an active hemicellulase and protease.

It was found that the solubility of dried grasses in *Trichoderma* cellulase to be highly correlated (r = 0.92) with in *vivo* digestibility. The crude cellulase used contained 71 g kg^{-1} protein and 640 g kg^{-1} of water-soluble carbohydrate, half in the form of reducing sugars. The enzyme preparation contained active cellulases, hemicellulases and proteases, and completely hydrolysed isolated cellulose and 70% of isolated hemicellulose from rye grass in 24 h; the cellulase removed about half the total protein in the grasses in 24 h. The amount of dry matter solubilized by the enzyme preparations used was 15-25 units lower than *in vivo* or *in vitro* (inoculum) values; a regression equation was therefore required to relate cellulase solubility to *in vivo* digestibility. It has been well established that cellulases from different fungal sources or from different suppliers can vary greatly in their capacity to solubilize native cellulose or plant cell. Cellulases from *Trichoderma* have generally been found to be the most effective and consistent. Some cellulases have no greater solubilizing activity than simple buffer solutions.

The effectiveness of *Trichoderma* cellulase is confirmed by other studies, showing simple solubility in cellulase to be well related to *in vitro* (inoculum) digestibility of grasses and legumes and to be applicable to the evaluation of fresh sorghum leaf and treated wood products. Other workers have studied the effects of combining cellulase digestion with subsequent treatment with other enzymes, notably proteases. The solubility in cellulase followed by pronase to be well related to *in vivo* digestibility for a range of forages, cob residues, straw, etc. Different regression equations were required for the different materials assayed, but, subject to this provision, cellulase solubility was considered a suitable alternative to inoculum methods. *Trichoderma* cellulase was found to digest forage cell walls more efficiently than *Aspergillus* cellulase; incubation with acid pepsin following cellulase treatment increased the amount of dry matter solubilized. Cellulase-pepsin solubility was highly correlated ($r = 0.87$) with *in vivo* digestibility, but different regression equations were required for grasses and legumes. *In vitro* (inoculum) digestibility predicted *in vivo* digestibility more precisely ($r = 0.95$) and was less influenced by species differences. Similarly, It is also concluded that *in vitro* (inoculum) digestion gave a better prediction of *in vivo* digestibility for tropical grasses and legumes than an enzyme method involving successive incubations with cellulase, amylase and pepsin.

3.8.2. The effect of different pretreatments on cellulase digestion

Incubation with acid pepsin

Incubation with pepsin in 0.1 mol 1^{-1} hydrochloric acid prior to cellulase treatment was found to increase the amount of dry matter solubilized and to significantly improve the correlation of enzyme solubility with the *in vivo* digestibility of grasses.

Incubation with pepsin after cellulase only slightly improved the correlation with *in vivo,* while pretreatment with 0.1 mol 1^{-1} hydrochloric acid had no effect on the correlation. It was concluded that acid-pepsin treatment improved the accessibility of the cell-wall polysaccharides to the cell-wall-degrading enzymes. Pepsin-cellulase solubility was also found to be highly correlated with *in vitro* (inoculum) digestibility for grasses ($r = 0.96$) and clover ($r = 0.94$), with very similar regression equations derived from the different species. The pepsin-cellulase method was found to be highly reproducible, with a coefficient of variation of 1.09% for successive batches of analyses. The technique has been in routine use at the Institute of Grassland and Environmental Research for the past 20 years. The only modifications to the original method described by Jones and Hayward (1975) are: (i) the reduction of incubation time in cellulase from 48 h to 24 h; and (ii) the use of a liquid *Trichoderma* cellulase (Celluclast 1.5 L from Novo Nordisk 4/5,2880 Bagsvaerd, Denmark). The liquid concentrate is diluted 10 $ml1^{-1}$ with the acetate buffer previously described and 20 ml used

200 mg "! of dried forage. The enzyme concentrate is purchased in 5 1 quantities and each batch calibrated against a set of samples of known *in vivo* digestibility. The regression equations used to predict digestibility are then adjusted, if necessary; changes in the equation over many years have been minor and would have little effect on predicted values.

Many workers have shown the pepsin-cellulase technique, as described by Jones and Hayward (1975) or with minor modifications, to predict *in vivo* digestibility with an accuracy comparable to inoculum methods. The method has been successfully applied to a variety of materials, including tropical grasses and legumes, grass silages, hays, treated straw and kale cell walls. Pepsin-cellulase solubility to be highly correlated (r = 0.94) with *in vivo* digestibility for both tropical and temperate grasses over a range of 40-76% in *vivo* digestibility. Using separate regressions for different grass species reduced the standard error of prediction. It was concluded that the enzyme method could accurately predict *in vivo* digestibility, provided samples of known in *vivo* digestibility similar to those being tested were included as standards in each run.

A simple digestion with cellulase to predict the digestibility of grasses as precisely as the pepsin-cellulase method; predicted digestibility for clover and high-protein feeds (peas, beans, fish-meal), however, was much improved by pepsin pretreatment. Pepsin pretreatment to improve correlations of cellulase solubility with in *vivo* digestibility for grain diets and forages; differences in substrate solubility by *Aspergillus* compared with *Trichoderma* cellulases were far less marked if samples were pepsin-treated. A cellulase from *Aspergillus* niger appeared to contain an amylase and was better suited for the analysis of cereal feeds than *Trichoderma* cellulase. Allison and Borzucki (1978) claimed that increasing the acid concentration for the pepsin stage and the incubation temperature to 50ªC gave pepsin-cellulase solubility values comparable to *in vitro* digestibility. Jones and Hayward (1982) compared this and other suggested modifications, but could find no improvement over the original procedure.

Terry *et al.* (1978) compared the pepsin-cellulase technique with the two-stage inoculum method (Tilley and Terry, 1963) for predicting *in vivo* digestibility for grasses, lucerne, red clover and sanfoin. They found the enzyme method to be very repeatable and only slightly less accurate than inoculum for predicting the digestibility of grasses (Table 8.4). Enzyme solubility was less accurate than the inoculum method, however, for predicting the digestibility of the legumes. For the inoculum method, a single regression equation was permissible for both grasses and legumes. Separate regressions were required for the grasses and for each legume species when the enzyme technique was used.

Aufrere and Michalet-Doreau (1988) have suggested an interesting variation of the pepsin-cellulase technique to cater for feeds containing starch. The method involves three stages: after the initial acid-pepsin incubation at 40ªC, the temperature is raised to 80ªC for 30 min to hydrolyse starch; the final stage is an incubation with cellulase after removal of the acid pepsin. The results using this technique were highly correlated ($r = 0.98$) with *in vivo* digestibility for a range of forage, straw and mixed diets, containing varying preparations of concentrates and byproducts.

Neutral-detergent extraction

Neutral-detergent extraction, under reflux or at 100°C, has been widely used as a pretreatment before incubation with cellulase.

Jones and Bailey (1974) showed cellulase digestion with neutral-detergent residues from grasses to be highly correlated with the voluntary intake of grasses by sheep. The procedure was considered appropriate for breeding programmes aimed at improving both digestibility and intake. Difficulties were encountered in routine use, however, due to the inhibition of cellulase digestion by traces of lauryl sulphate from the neutral-detergent reagent. The optical density (324 nm) of cellulase extracts of neutral-detergent residues from grasses was highly correlated with *in vivo* digestibility; the ultraviolet (UV) absorption was attributed to the release of carbohydrate esters of phenolic acids by the cellulase. Roughan and Holland (1977) prepared their own potent crude cellulase from a mutant strain of *Trichoderma,* which readily attacked plant cell wall isolated by neutral-detergent extraction. They proposed a method involving hot neutral detergent, followed by exhaustive digestion with two successive batches of cellulase. Values obtained using this technique were highly correlated ($r = 0.98$) with *in vivo* digestibility of legumes and grasses. Comparison of cellulase solubility, using varying concentrations of cellulase was made, after either pepsin or neutral-detergent pretreatment, with in *vitro* (inoculum) and in *vivo* digestibility for a range of forages, silages, cereals and complete diets. An additional digestion with amyloglucosidase was included in the enzyme methods for materials containing starch. Some results are summarized and show the enzyme techniques to be at least as precise as *in vitro* inoculum for predicting the *in vivo* digestibility of most forages; there was little difference found between the two enzyme methods. The authors point out that the pepsin-cellulase method was easier to manipulate and less time-consuming than the detergent method; an advantage of the latter was that samples could be analysed in fewer working days. It was concluded that both cellulase methods could be regarded as acceptable alternatives to the *in vitro* inoculum method for the routine evaluation of forages.

Modified detergent-cellulase methods have also been proposed to accommodate the galactomannans found in palm-kernel oil. These polysaccharides are not solubilized by cellulase but can be solubilized by the inclusion of an additional carbohydrase, gamanase, in the buffered cellulase and an increase in digestion time to 40 h. The modified method significantly improved the correlation with *in vivo* metabolizable energy values; the inclusion of gamanase did not affect predicted energy values for feedingstuffs not containing palm-kernel oil. It was concluded that the extra expense of the modified method was justified for compound feedingstuffs containing 10% or more of palm-kernel oil; for samples known to be free of palm oil, the original detergent-cellulase method would remain valid. Downman (1993) points out that new commodities introduced into the feed trade in the future will need to be monitored to decide on the appropriate analytical methodology required. A routine procedure for the determination of neutral detergent cellulase-gamanase solubility (NCGD) has been described by the Ministry of Agriculture, Fisheries and Food (MAFF)-sponsored working party on the prediction of energy values of compound feeds (MAFF, 1993). Appropriate test kits have been prepared for the technique (Biotal Ltd., Cardiff) and the working party specify that kits from this source must be used in support of their prediction equation for the metabolizable energy content of ruminant compound feeds.

Comparison of the accuracy was made for the detergent and pepsin-cellulase methods for predicting the *in vivo* digestibility of a range of grass and legumes. Both cellulase methods were highly correlated with *in vivo* digestibility values; pepsin pretreatment appeared to be more precise for grasses and detergent pretreatments for legumes. It was concluded that pretreatment with detergent increased analytical error and that the method failed to produce an overall improvement in the accuracy of predicting *in vivo* digestibility; any advantage gained from a shorter analysis time was offset by an increase in analytical error. Similarly, some workers found no advantage of detergent, compared with pepsin, pretreatment for predicting *in vivo* digestibility for lucerne (alfalfa), a range of grasses and ryegrass-lucerne mixtures. The addition of *Aspergillus* hemicellulase to *Trichoderma* cellulase improved the correlation with *in vivo* values for both pepsin and detergent methods. The use of this enzyme combination following pepsin pretreatment predicted *in vivo* digestibility as precisely as the two-stage *in vitro* (inoculum) method studied (r^2 = 0.88).

A simple digestion in buffered cellulase to be highly correlated (r = 0.91) with the *in vivo* digestibility of a range of temperate forage species was noticed. The correlation with *in vivo* was improved, however, if the forage was pretreated with neutral detergent (r = 0.97) or acid pepsin (r = 0.94) prior to cellulase digestion. They concluded that the one-stage cellulase technique would be adequate as a rapid, low-cost, method for screening in plant-breeding programmes. More precise estimation required pretreatment with detergent or pepsin; both enzyme methods were considered to be more rapid and convenient and less costly methods of predicting *in vivo* digestibility than the two-stage *in vitro* inoculum method of Tilley and Terry (1963). Subsequent work ers showed similar results for lucerne. Again, simple digestion in cellulase was well related to in *vivo* digestibility (r = 0.91), but the relationship was improved by pretreatment with detergent (r = 0.94) or acid pepsin (r = 0.96). Similarly, both detergent- and pepsin-cellulase methods appeared to be as precise as *in vitro* inoculum for predicting the *in vivo* digestibility of fresh grass for dairy cows.

Hot-acid extraction

More drastic pretreatments than pepsin or neutral detergent have been proposed in efforts to increase the amount of dry matter solubilized by cellulase to levels comparable with *in vivo* digestion. The refluxing with 2 mol 1^{-1} hydrochloric acid for 30 min, followed by digestion with cellulase and then pepsin, to give values numerically similar to *in vivo* digestion for a range of forages (grasses, legumes and cereal straw). Results obtained with this method were highly correlated (r^2 = 0.85) with *in vivo* digestibility and the method was concluded to be a viable alternative to inoculum methods. Few comparisons appear to have been made between the acid-reflux method and the detergent-cellulase or pepsin-cellulase procedures. *In vitro* (inoculum) values for grasses and legumes to be highly correlated with pepsin-cellulase solubility (r = 0.96) and the reflux-acid method (r = 0.91). Different regression equations were needed to predict *in vitro* values of grasses and legumes for pepsin-cellulase but not for the acid-reflux procedure. Highly significant week-to-week differences were found between different analytical runs for the inoculum procedure, but not for the enzyme methods. The reflux-acid method compared favourably with the detergent and pepsin methods when applied to lucerne forages.

Comparison and the accuracy of different enzyme techniques and *in vitro* inoculum for predicting the *in vivo* digestibility of maize silage, grass silage and grass hays was studied. The enzyme techniques included: (i) the 2 mol 1^{-1} acidpepsin-cellulase procedure; (ii) pepsin-cellulase solubility with varying incubation times for the pepsin and cellulase stages; and (iii) a variant of the pepsin-cellulase method involving three stages, namely: incubation in acid pepsin, heating the acid pepsin solution at 80°C for 45 min to remove starch and finally incubation with cellulase. The detergent-cellulase method was not evaluated, as this method had proved more difficult to manipulate than pepsin-cellulase, resulting in a lower repeatability. The *in vitro* inoculum method was the best predictor of *in vivo* digestibility for grass silage and hay, but the enzyme methods were generally superior for maize silage, particularly methods involving a hot-acid pretreatment. No one enzyme method was best for all forages, but the pepsin-cellulasebased methods were more consistent than those involving acid-reflux. A similar method to that proposed but with a shorter heating time at 80°C, has been suggested. The method was found to be more precise than the *in vitro* (inoculum) procedure for predicting the *in vivo* digestibility of a diverse range of feeds, including cereals, byproducts and dried forages.

References

Aufrere, J. and Michalct- Doreau, B. (1988) Comparison of methods for predicting digestibility of feeds. *Animal Feed Science and Technology* 20, 203-218.

Donefer, E., Niemann, P.J., Crampton, E.W. and Lloyd, L.E. (1963) Dry matter disappearance by enzyme and aqueous solutions to predict the nutritive value of forages. *Journal* of *Dairy Science* 46, 965-970.

Downman, M.G. (1993) Modifications to the neutral detergent cellulase digestibility method for the prediction of the metabolizable energy of compound feeds containing palm kernel oil. *Journal* of *the Science* of *Food and Agriculture* 61, 327-331.

Downman, M.G. and Collins, EC. (1982) The use of enzymes to predict the digestibility of animal feeds. *Journal* of *the Science* of *Food and Agriculture* 33, 689-696.

Jones, D.1.H. and Bailey, R.W. (1974) Hydrolysis of cell wall carbohydrates of grasses by carbohydrases in relation to voluntary intake by sheep. *Journal* of *Agricultural Science, Cambridge* 83, 105-111.

Jones, D.1.H. and Hayward, M.V. (1982) Estimation offorage digestibility. In: *Annual Report Welsh Plant Breeding Station* for 1982, pp. 126-127.

MAFF (1993) *Prediction* of *the Energy Values* of *Compound Feedingstuffs for Farm Animals. .Summary* of *the Recommendations* of *a Working Party Sponsored by the Ministry* of *Agriculture, Fisheries and Food.* MAFF Publications, London, 17 pp.

Roughan, P.G. and Holland, R. (1977) Predicting in *vivo* digestibilities of herbages by exhaustive enzyme hydrolysis of cell walls. *Journal* of *the Science* of *Food and Agriculture* 28, 1057-1064.

Terry, R.A, Mundell, D.C. and Osbourn, D. (1978) Comparison of two in *vitro* procedures using rumen liquor-pepsin or pepsin-cellulase for prediction of forage digestibility. *Journal* of *the British Grassland Society* 33, 13-18.

Tilley, J.M.A and Terry, R.A (1963) A two-stage technique for the in *vitro* digestion of forage crops. *Journal* of *the British Grassland Society* 18, 104-111.

van Soest, P.J. (1963a) Use of detergents in the analysis of fibrous feeds. I. Preparation of fibre residues of low nitrogen content. *Journal* of *the Association* of *Official Agricultural Chemists* 46, 825-829.

van Soest, P.J. (1963b) Use of detergents in the analysis of fibrous feeds. II. A rapid method for the determination of fiber and lignin. *Journal* of *the Association* of *Official Agricultural Chemists* 46, 829-835.

□□□

Chapter 4

Techniques in Grazing Experiments

4.1 Estimation of Herbage Yield of Pasture

Herbage yield (kg ha⁻) is a measurement essential to interpret the response of herbage availability on animal performance. A single measurement provides an instantaneous estimate of herbage, multiple measurements at regular intervals track the changes that occurs during a grazing period or for a growing season. Herbage yield provides a guide to pasture conditions and permits stocking of animals as per the carrying capacity of pasture. It also provides better understanding of the animal response with varying herbage yield.

The most accurate way to measure herbage yield is by harvesting the total plot at ground level. This is possible only in small plots. In grazing experiments, plot size is very large and maximum herbage mass must be left for grazing. Both of these methods require sub-sampling.

Methods for estimating herbage yield can be classified as direct or indirect.

Direct Method

Direct method involves harvesting of herbage from a quadrate of specific frame size at ground level and yield is expressed on a dry matter basis.

1. Quadrate method
2. Cage method

Indirect Method

All indirect methods require calibration when quantification of herbage mass ($kg\ ha^{-1}$) is desired. The calibration employs a double sampling technique, in which indirect reading is associated with actual herbage mass obtained by direct sampling. The relationship between the two is used as the basis for predicting direct value.

1. Calibrated weight estimate method
2. Rising plate method
3. Capacitance probe method

Quadrate method

It is a conventional and simple method of estimating herbage yield. The number, type, size and location of the quadrate is decided, depending upon the accuracy of the estimate needed. Herbage yield is reported on a dry matter or organic matter basis, depending on degree of soil contamination.

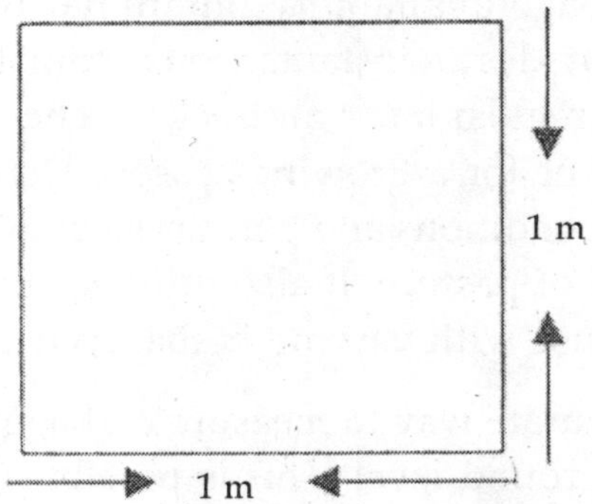

Fig. 1: Iron rod quadrate for estimating herbage yield

Procedure

The quadrate is made up of wire, iron pipe or rod or wooden frame of 0.25 to $1m^2$ size. The sampling quadrate is placed at random in the pasture plot so that the differences due to plant density, height and composition in the plot are avoided. The sampling frame is placed at a constant pace intervals along the linear transects. The number of quadrate per hectare must be six. Herbage within the quadrate is clipped 2 cm above the ground level at uniform height. Herbage is weighed in the field on a spring scale and placed in a paper bag with quadrate number and estimated weight marked on . it. The samples are subsequently dried in an oven at 70-80°C for 24 hr and yield is expressed on dry matter basis.

Advantages

1. Method is very simple and practicable, does not require sophisticated instruments
2. Inexperienced persons can also adopt this technique.

Disadvantages

1. In sparse and patchily distributed vegetation cover. skilled decision is required to avoid any bias in estimating herbage yield.
2. In instances where, many plants are rooted within the quadrate, but its most of the foliage remained out side the quadrate, then whether it is pulled back in to quadrate or not? Adopting a consistent procedure for all the quadrates can minimize this problem.

Further reading

Papanastasis P V.1977. Optimum size and shape of quadrate for sampling herbage weight in grassland of Northern Greece. Journal of Range Management 30: 446-448.

Cage method

The measurement of herbage growth usine, an enclosure is known as cage method. The cage method is used to estimate herbage yield and the amount of herbage consumed by the animals grazing a pasture plot dominated by annual species. Cages may be either placed in one position for whole season, known as permanent cages or moved to different positions several times during a season, known as movable cages.

Procedure

The size of lx1 m cage is randomly placed in the pasture area of 1 ha nearly on 6 sites. At each sampling, the annual and biannual plants inside the cage are clipped to approximately 2cm above the ground level. The samples of each quadrate are dried at 70-80°C for 24 hr and weighed. Herbage yield is estimated from total herbage mass harvested from the cage.

Herbage yield (kg/ha) =Herbage mass of a cage (kg) x 10,000

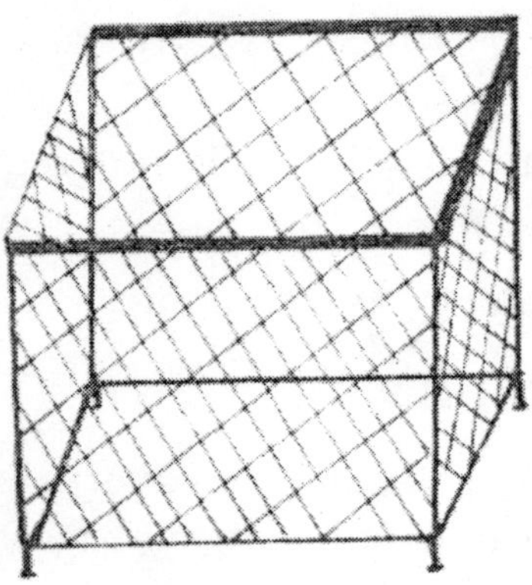

Fig. 2 : Empty Cage

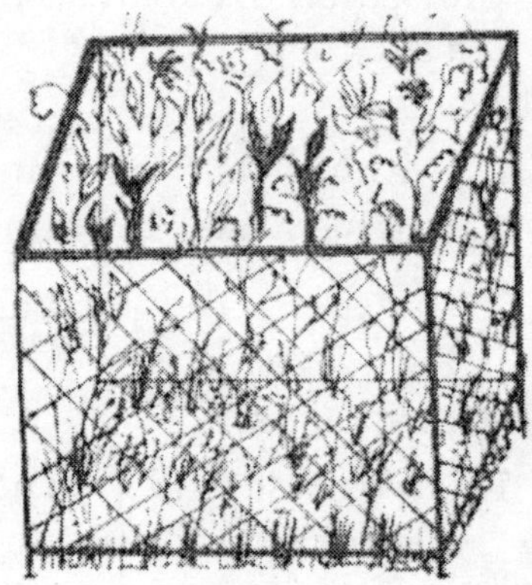

Fig. 3 : Cage with herbage growth

Advantages

1. Cage method gives accurate estimate of herbage mass for a specific period.
2. This method is suitable to annual and biannual plant species.

Disadvantages

1. In heterogeneous and patchily distributed vegetation, cage method gives wide variability in herbage yield estimation.
2. Cage method is found unsuitable for short spell vegetation species.
3. When used as a method to estimate herbage consumed by the animal on a pasture, it often over estimates the value.

Further reading

Zahran B B, Holm A Mcr, Stern W R and Loneragon W A.1988. The cage method of sampling to assess herbage mass and herbage consumed on annual rangeland vegetation. Australian Rangeland Journal. 10: 6-12.

Calibrated weight estimate method

The method is more useful, when there is physical limitation of cutting large number of samples, and the need to ensure that the area is not sufficiently large to become a significant treatment. Under such a situation, a double sampling technique of visual weight estimates calibrated by harvesting is applied for the measurement of herbage yield.

Procedure

A quadrate (made of wire or iron pipe or wood) of 1m² is placed at a constant pace intervals along the linear transects. An observer is trained to calibrate the visual estimate of herbage mass from a quadrate, to those obtained by harvesting to the extent that both the values are comparable. A pre-determined proportion of estimated quadrate, preferably every fifth is uniformly clipped 2cm above the ground level immediately after its visual observation and weighed on a spring scale in the field. The material is stored in a paper bag with quadrate number and estimate weight marked on it. In each field the ratio of number of visual observation quadrates to that of harvest one should be 4:1.eg. 24 of them by visual observation and 6 of them by harvesting.

Calibration

Multiple values of visual estimates of fresh yield and their corresponding dry yields are recorded. Then a regression equation Y=a+bX is developed, *where* Y =- actual dry weight and X = visual estimates of fresh yield.

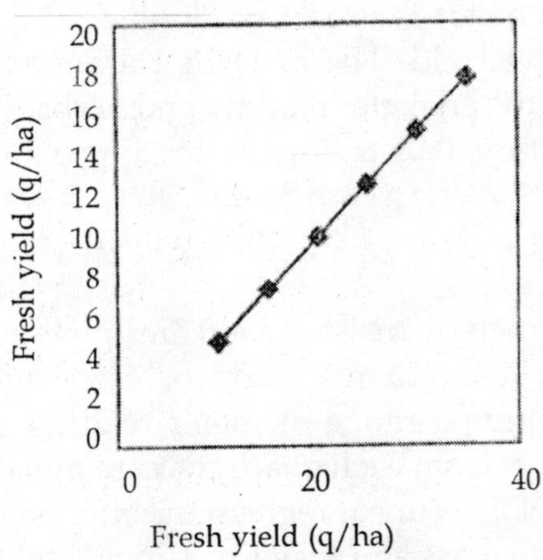

Fig. 4 : Relationship between fresh and dry yield of herbage

Then, by substituting different values of fresh yield, the corresponding values of dry yield are obtained. A regression line between two values is plotted. This equation can be used to calculate the value of dry yield from known value of estimated fresh yield.

Advantages

1. More suitable for small plots
2. Large number of samples can be obtained without actually harvesting, them.

Disadvantages

1. In-experienced person with lack of natural skill may bias the results
2. Method is not suitable for heterogeneous pasture vegetation with varying phenlogical stages
3. The consistency of estimate decreases, when vegetation is stemmy/leafy, prostrate/ erect and broad/narrow.

Further reading

Tadmor N H, Beight A, Noy Meir R W and Eyal E.1975. An evaluation of calibrated weight estimated method for measuring production in annual vegetation. Journal of Range Management. 28: 65-69.

Rising plate meter method

A rising, plate meter is used in a double sampling technique to measure the herbage yield. The meter is calibrated by developing a linear regression between meter reading and herbage dry matter mass, as measured by cutting, 0.25 tri- quadrates approximately 2 cm above the ground *level*. The meter is quick, easy to

Procedure

The Ellinbank meter consists of a $0.3m^2$ plate with a downward pressure of $4.5kg\ m^{-2}$. Six $0.25\ m^2$ quadrates of herbage of known meter reading are cut to give a range of meter reading in each paddock. While cutting herbage from each quadrate, care must be taken to avoid contamination with soil. A linear regression equation is developed from the values of herbage mass and meter readings. Then, herbage mass is calculated from the meter readinu, by using the formula.

$$Y = bH + a$$

Where,

Y = herbage mass (kg DM ha^{-1}) and

H = meter reading (cm)

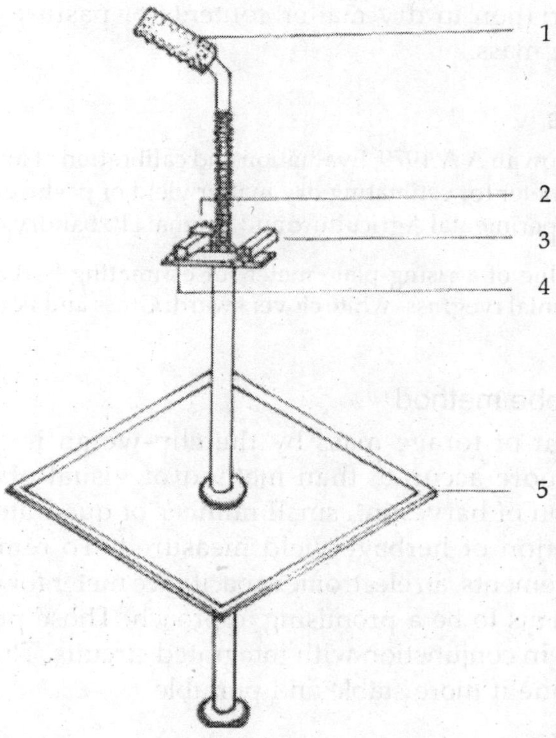

Fig. 5 : The Ellinbank pasture meter

1. Handlebar grip 2. Revolution teeth 3.Number recorder 4. Height recorder 5. Rising plate

Advantages

1. Little time and efforts are required for estimating the herbage mass
2. It is not necessary to cut herbage quadrate at each sampling date, because it follows distinct and repeatable calibration relationship throughout the year.
3. It produces a two-term calibration reading and this will overcome the problems of cutting the herbage at uniform height at ground level.

4. Linear increase in meter reading provides corresponding actual estimate of herbage mass.

Disadvantages

1. Accumulation of dead litter on ground is difficult to account
2. Diurnal variation in dry matter contents of pasture often bias the herbage mass.

Further readings

Earle D F and Mc Gowan A A.1979. Evaluation and calibration of an automated rising plate meter for estimating dry matter yield of pasture. Australian Journal of Experimental Agriculture and Animal Htzbandry. 19: 337-343.

Michall P.1982. Value of a rising-plate meter for estimating herbage mass of grazed perennial ryegrass- white clover sward. Grass and Forage Science 37: 81-87.

Capacitance probe method

Measurement of forage mass by the clip-weigh is costly and tedious, but is more accurate than method of visual observation. Physical limitation of harvesting, small number of quadrates disturbs the actual reflection of herbage yield measured. To remove these conflicting requirements, an electronic capacitance meter for estimating herbage yield seems to be a promising approach. Those probes have single rod sensor in conjunction with integrated circuits. These special features have made it more stable and portable.

Procedure

A probe is randomly placed at regular spacing (e.g. every second step) during a systematic walk over each plot. Air readings are taken frequently. The air readings are subtracted from the pasture readings to obtain corrected meter readings (CMR). A CMR registers, the increase in capacitance, when the probe is placed on the pasture and thus reflects the herbage mass in the area surrounding the probe.

Calibration

For calibration, intensive probe readings are taken within 1 m^2. Twenty-four readings per m' in creeping type and 48-readings perlm- in erect type of vegetation are taken. These are then averaged. The herbage within the four random chosen quadrates are cut 2 cm above the ground level by shear and weighed by spring scale.

A linear relationship between herbage mass and corrected meter reading is established.

HM= a+b CMR

is used to estimate herbage mass (HM) from corrected meter reading (CMR).

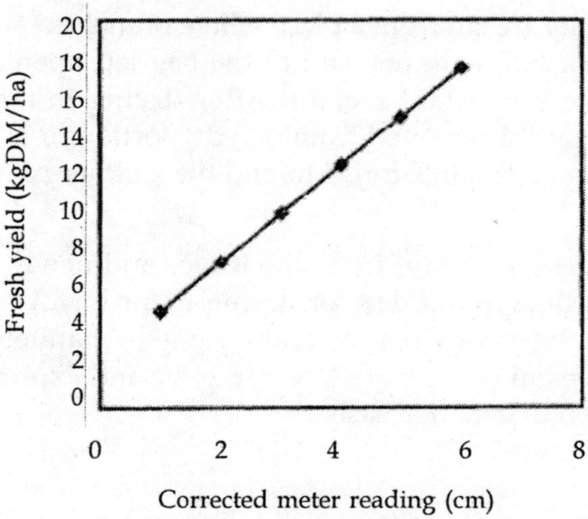

Fig. 6 : Relationship between corrected meter reading and herbage mass

Advantages

1. This method precisely estimates the herbage mass
2. Saves time and effort compared to quadrate method.

Disadvantages

1. More frequent waxing of probe under wet conditions is required for accurate estimation of' herbage mass.
2. Requires frequent reading of air when atmospheric conditions are changing rapidly.

Further reading

Csosbic S F. Smallfield B M. Hawker H, Floate M J S. Keoghan J M. Enright P D and Abernethy R J. 1987. Exploiting the pasture capacitance probe in agricultural research: a comparison with other methods of measuring herbage mass. Journal of Agricultural Science, Cambridge. 108: 155-163

4.2 Estimation of Shrub and Tree Foliage Yield

Total foliage production from shrubs or trees is one of the most difficult parameters to estimate on native rangeland. There are several methods for estimation, but the adoption of method depends upon its efficiency in terms of time saving, non-descriptiveness and preciseness.

Cut and weigh method

Shrub plants are cut from a place where branches start and placed in a large paper bag with one end of the bag left open. The samples arc air dried to a constant weight. After drying in air. leaves and ephemeral twigs are removed from woody portion. It is sub-sampled for oven drying at 70- 80°C for 24 hr and the yield is expressed on dry basis.

While in case of fodder trees, the leaves and newly grown twigs are lopped on the ground, left for drying in air for 24 hr. Leaves are removed from the woody portion and weighed separately. Keep small portion of leaves in oven at 70-80°C for 24 hr and express the foliage yield of trees on dry matter basis.

Advantage

It provides accurate estimates of foliage yield.

Disadvantages

1. Requires more labour and cost, in comparison to other methods.
2. Research design does not suit to adopt this method in most of the animal experiment
3. 1 h is difficult to take more number of samples to make treatment statistically significant, because sampling greatly reduces the availability of foliage to animals.

Further reading

Wilson A.D.Abraham N A. Barran R. Choate J. Green D R. Harland R J. Oxley .R E and Stanley R J.1987. Evaluation of method of assessing vegetation changes in the semi-arid rangeland of Southern Australia. Australian Rangeland Journal. 9:5-13.

4.2.1.Dimensional method

This method works on the principle of relationship between height and weight to foliage yield. However, every individual species has different level of relationship between these two parameters. Moreover, in experiments-here more number of samples are. required to make the tratment effective, this method is more useful.

Procedure

Total number of shrub plants of one species should be 30-35 at one site to make the coefficient of determination (R2) more accurate. Each plant is marked and 21k-4ted a number for identification. Two diameters (the 1.-vages: and perpendicular to the longest) and the foliage Nicreit are measured in each plant. After the measurement. me fifth of total number of plants are clipped to ground kvel by hand shear and total foliage mass is estimated.

For inverted conical growth plant, canopy volume is determined by,

Canopy volume =1/3 m r2 h

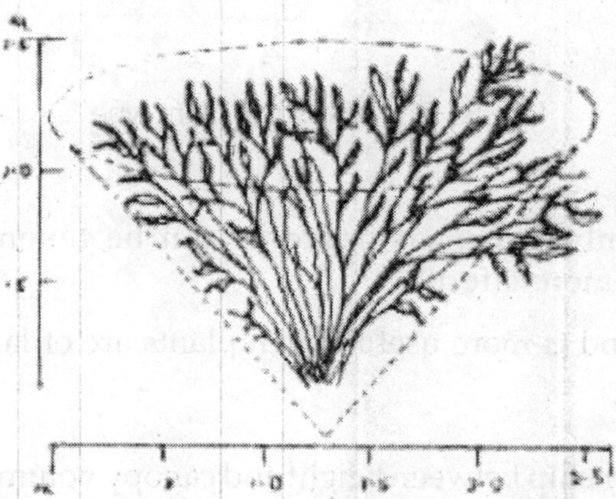

Fig. 7 : The inverted conical plant type

For cylindrical type of plant,

Canopy volume = $\pi r^2 h$

Where, r = radius and h = height

The relationship between the actual weight and canopy volume is established using a simple linear regression. If curvilinearity is apparent, log transformation and polynomial mode are used to obtain the best fit. The preferred equations are selected on the basis of coefficient of determination (R2).

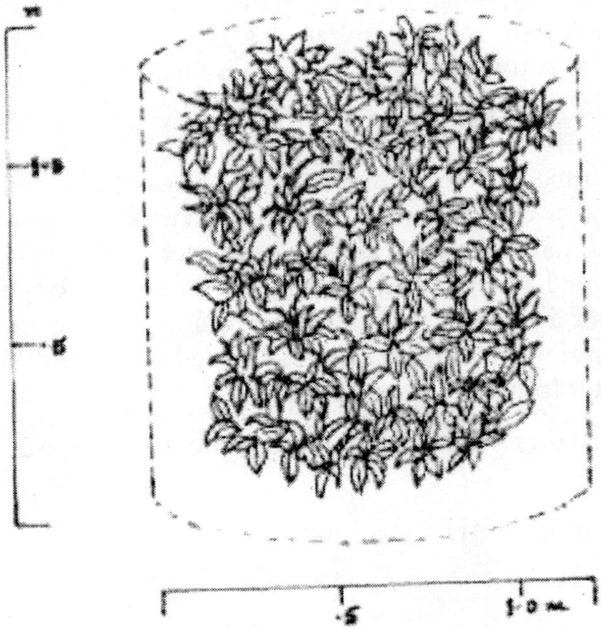

Fig. 8 : The cylindrical plant type

Advantages

1. More number of measurements can be taken to make the treatment more effective.
2. This method is more useful when plants are of large size.

Disadvantages

1. The relationship between height and canopy volume of individual plant species varies with the phenological stages and seasons. Thus relationship once established for individual species is not valid for other season or age.

Further reading

Kirmse R D and Norton B E.1985. Comparison of the reference unit method and dimensional analysis methods for two large shrubby species in the Coatinga wood land. Journal of Range Management. 38: 425-427.

4.3 Estimation of Basal Cover of Pasture Plot

Foliar cover of the herbaceous layer and the canopy cover of the shrub-tree species are determined by a sample of points along a transect.

Number of points and transect length and width is decided depending upon plot size and experimental design. The points along a transect are located by wheel point or step point. The foliage canopy of the plant is defined by its perimeter ignoring gaps in that canopy. Only the upper most strike is recorded for herbage chenopod shrub layer and lower most strike for tall shrub- tree layer, counting each layer separately. This method records the shrubs, trees and dominated herbaceous to the species level but other minor species, in particular annual are identified only to a group level. Estimated sum of basal cover of individual species including, herbaceous, tree and shrubs provide the basal cover of the area.

Procedure

All individual plants in a quadrate of 1m- size are counted, harvested to ground level with a shear and placed in a paper bag. The basal cover of individual harvested plant species is estimated by unmarked over lay. The unmarked Perspex sheet (15 x15 cm) is placed on each of the clipped plant species. The periphery of each plant is outlined, delineated and transformed to a sheet of white paper. The herbage basal cover is then obtained by taking the mean of individual species: This estimates the basal cover of a species/1m2. Basal cover of individual species is then expressed as percentage of total cover. Basal cover of a plant species in terms of in ha I is obtained by multiplication of' basal cover of individual species with number of plants of same species.

Disadvantages

1. The empty space in the canopy overestimates the value.
2. The canopy cover of minor species is estimated only to group level, not to species level.

4.4 Botanical Composition of Pasture Plot

The botanical composition of grassland vegetation is a reflection of many factors, including past management. Changes in these factors will be reflected in changes in composition. There are many attributes of botanical composition that can be described or measured and the objective of the description or measurement dictates the methodology. Monitoring changes over time by repeated sampling is a commonly used approach to detect environmental or management effects.

Vegetation description or measurement can be very labour intensive and time consuming. The resources invested depend on the purposes of the measurements and, essentially, the information collected is usually proportional to the resources invested but there is wide variation in the cost effectiveness of the different methods available. Scale is also important and different methods must obviously be used for vegetation measurement on a global scale compared with a small replicated experiment.

This chapter provides an overview of the methodologies available for the measurement of the botanical composition of grassland vegetation appropriate to a range of different objectives.

An understanding of the composition and quality changes of pasture throughout the year helps in formulation of range management plans to optimize the use of forage resources. Rangeland broadly consists of shrubs, forbs and grasses and together contributes to total herbage yield. Wide seasonal changes in distribution of grasses, forbs and shrubs are noticed in tropics. Generally grasses are more frequently distributed in monsoon and early winter and withers off with the on set of summer. Fallen mass of forbs, grasses and shrubs form (lead litter, which also sizably contributes to availability of feed resources during dry season. Accurate measurement of individual plant species contribution in the grazing land in terms of dry matter yield as well as frequency is required. Moreover there is a need to identify palatable and unpalatable species in the ranges, so that, unpalatable plant species could be eliminated, to improve the yield and quality of ranges.

Scales of Measurement

Different scales of measurement of grassland vegetation require different levels of detail. Satellite technology has made it possible to distinguish grasslands from other forms of vegetation and to measure certain attributes at the global scale but there is still a gulf between the detail possible with remote sensing and that possible on the ground. There are two separate approaches when moving between scales. The first is where very detailed data are collected and then aggregated to give a big picture. The second is to collect data on a broad scale and then focus in on the detail required for a particular purpose.

Aggregation of detailed data using geographic information system (GIS) technology is now well developed and there are no fundamental technical problems in moving from very detailed data collected in small plots to a global scale, provided sufficient data, collected in a comparable way, are available. However, the technology to move in the other direction in order to see the fine detail in a small area, using data collected at the global scale, is still a long way away.

Global scale

Satellites can provide data over the whole of the globe within a 24 h period. Using appropriate reflectance bands grasslands can be distinguished from other vegetation types and the rate of growth at low vegetation densities, ground cover and broad groups of species can be distinguished. Because the method of data collection is uniform over the whole world, direct comparisons can be made among different grasslands across international boundaries and over time or of the effects of different management or different seasonal conditions.

GIS technology involves digitizing a map of a region and then using computer technology to overlay detailed geographical data, which can include species composition. Detailed botanical composition data can then be aggregated to produce broader-scale maps, provided the requisite base maps and the detailed botanical composition data are available. In practice, however, there are several constraints to the use of GIS technology for this purpose. Until compatible digitized base maps are available for the whole world it is not possible to aggregate local data on a world-scale GIS database. Botanical composition of grasslands across regions must be collected in the same way for the results to be comparable across national and international boundaries. Uniform data compatible with GIS databases are simply not yet available for many of the world's grasslands and so valid aggregation to give a global picture is not yet possible.

National and regional scale

Many national and regional maps of vegetation, including grasslands are available. Such maps were mostly produced by ground surveys, usually by road, with periodic stops to list floristics. The surveyors used a broad vegetation classification system and simply marked the boundaries of different communities on a base map. Species lists were usually provided for the different communities with some estimates of the relative abundance of the different species. These maps have often been built up over many years, with subsequent authors correcting and improving previously published maps. One problem is that the mapping units sometimes change, and problems of nomenclature arise with taxonomic changes. This subjective classification of vegetation, with somewhat randomly collected species lists at regular or irregular intervals across the landscape, has now largely been superseded by more objective methods of site selection, quantitative data collection and data analyses. These objective methods will be dealt with in more detail later. However, there may still be some parts of the world where the subjective approach represents the only available description of national and regional grasslands.

The more detailed the data collection, the more intensively the vegetation can be sampled for species composition and classification of grasslands. In many countries black-and-white or colour aerial photographs are commercially available, either as mosaics or as overlapping individual frames that can be viewed with a stereoscope to obtain a three-dimensional perspective. The aerial photographs can then be used to select sites or transects for the ground collection of descriptive or quantitative botanical composition data, depending on the purpose of the measurements. The aerial photographs and accompanying ground-survey botanical composition data can be digitized and incorporated into a database using GIS software.

Farm scale

The simplest approach to obtaining botanical information of grasslands on a farm scale is to walk over the farm and record the species encountered with notes on their relative abundance. Depending on the size ofthe farm, it may be possible to record species over the whole area. For larger areas, the examination of sample sites is essential, with subsequent classification and aggregation of the data into some form of map. These maps can be valuable for planning farm operations. Searches for threatened grassland species and the identification of grassland communities with high conservation value are becoming increasingly important on farms. These species and communities are best located by either systematic or random walks, with the operator using prior knowledge of where such species and communities are most likely to occur in the landscape.

Extension services and farmers often use aerial photo mosaics as a basis for farm planning. Increasingly, with the ready availability of personal computers, a farm map can be digitized and many kinds of data, from botanical composition and pasture condition to livestock movements, included in farm GIS software. Extension services in many places provide user-friendly software so that farmers can modify and update their own GIS data for farm planning and farm record purposes.

Satellite imagery at the farm scale can be incorporated into such GIS databases. However, satellite data are expensive and special skills and equipment are necessary for their interpretation and incorporation into GIS databases.

Paddock scale

Most of the methods used to estimate botanical composition are designed for the paddock scale. This also represents the limit of the usefulness of satellite imagery, depending on the size of the paddock. Individual pixels can be located on the ground and the botanical composition related to reflectance data on specific wave bands, which can then be extrapolated to other parts of the paddock or further afield. Paddocks are often the experimental units in grassland management research where detailed botanical composition data are essential.

Many farmers are well aware of the dominant species in the different paddocks on their farms. Increasingly, there is a need for farmers to have additional information about the presence of other species, such as weeds or grassland plants of high conservation value, that may have an impact on paddock management.

Patch scale

Grasslands are almost always patchy except for newly sown pastures where the individual plants may be regularly spaced because of the method of sowing. Patches can vary in size from a few centimetres to more than 100 m across, and the smaller patches are too fine for detection by satellite imagery, though aerial photography can be very useful.

Patches can result from dung and urine voidance, in which case they may be temporary, whereas selective grazing by wild or domestic animals can produce long-term patches. The development of these patches is sometimes an early sign of weed invasion or impending land degradation and so their recognition and study become very important. Permanent patches can result from differences in soils or microtopography and position in the landscape. The interactions among topography, water movement, soil water availability and grazing animals affect the functioning of ecosystems and consequently the botanical composition in different parts of the landscape. Patches of high soil fertility can result from the grazing and defecation behaviour of domestic animals. These patches are long lived and have a different species composition than patches with lower fertility.

Measurement of botanical composition at the patch scale is becoming increasingly important in grassland management studies. The study of patches may include plant density or frequency measurements and patch dynamics within paddocks. It may involve the study of the population biology of individual grassland species or functional groups of species.

Sampling

Some form of sampling is always involved in measuring the botanical composition of grasslands. The number, size and location of sampling units and the timing of the sampling are critical. In small plot experiments, the whole population can sometimes be measured. The errors inherent in sampling vary enormously, depending not only on the spatial and temporal variation of the grassland attributes being measured but also on observer bias, and within and between the various sampling techniques. It is desirable to consult a statistician during the planning stages of any project involving grassland sampling to ensure that the sampling strategy is appropriate for the purpose and is statistically sound.

Sampling strategies

Suitable strategies depend on the characteristics of the grassland and the purposes of the measurements. The subsequent statistical analysis of the data is also governed by how samples are located in space and time.

The distribution of individuals of the species present in the grassland can affect the sampling strategy and the subsequent statistical analysis of the results. Non-random distributions are common and may confound parametric methods of statistical analysis. In contrast, the dry-weight-rank (DWR) method of botanical analysis of grasslands may not be appropriate on grasslands with random distributions. It may therefore be desirable to determine the type of distribution that occurs before deciding on sampling strategy, sample size and the statistical analysis of the results. Some species tend to be clustered while others tend to be regularly dispersed.

The distribution of species is likely to change with time, newly sown pastures having a more regular distribution of species than older pastures. It is also likely that the distribution of some species will vary over time and space with management and rainfall patterns. The nature and magnitude of non-random distributions also depends upon the scale at which the distribution is sampled.

Whether or not the distribution of an individual species in the community is random can be readily determined by observing a number of sample areas (quadrats) (see Kershaw, 1964, for details). Quadrat size should be chosen so that about three-quarters of the quadrats will contain no individuals of the species being measured. These quadrats are randomly located within the sample area and the number of quadrats containing 0, 1, 2 or 3 etc. individuals of the species are counted. The mean density is then calculated. The number of quadrats containing 0, 1,2 or 3 etc. individuals will follow a Poisson distribution if the individuals are randomly distributed and can be tested using chi squared for the goodness of fit. If it does follow a Poisson distribution, then the mean: variance ratio should not deviate significantly from 1.

Random sample location

Random sample location is appropriate where random spatial variation occurs in the grassland vegetation and the aim is to 'average out' this variation across the whole of the sample area. Random sample location is suitable for experiments where different treatments have been applied to different parts of an area of grassland in some form of replicated experiment. Choosing sample locations seemingly at random by the operator requires caution to avoid unconscious bias in their choice. Predetermined procedures should be used: for example, the tip of the left or right boot after a predetermined number of steps or, when an obstacle is encountered, a certain number of steps sideways to the left or to the right. One disadvantage of this method is that the operator unconsciously tends to avoid areas of prickly and sharp vegetation. Grid coordinates for sampling points chosen according to a list of predetermined random numbers can be located by pacing or by the use of a tape or a knotted rope, though their use may be too time consuming. Fence posts can be used in lieu of grid references in small paddocks.

Throwing a quadrat for locating sampling sites, or some other object such as a dart with a streamer attached, is not recommended because of unconscious aiming. Furthermore, taller or stemmy patches of vegetation are likely to intercept the quadrat and result in a biased distribution.

Another sampling strategy is 'systematic random sampling'. The first sampling point is chosen at random and then subsequent locations are selected at regular intervals by pacing or at set intervals along a tape or knotted rope. For repeated sampling to detect changes in the species composition over time, a knotted rope can be stretched between two permanent markers and the quadrats located at the knots for each sampling. Strictly speaking, the sample points are not independent of one another but, provided the points are spaced widely enough so that the same plant is not sampled at two consecutive points, the sample placement is essentially random.

Fixed grid sampling

A fixed grid sample layout can be used to detect vegetation patterns within an area of grassland. This method is essentially 'systematic sampling', because the initial points of the grid are not randomly placed. The location of each sampling point needs to be precisely fixed and the different grassland communities identified and mapped. Temporal changes in these patterns can be detected by repeated measurements using the same grid locations. Serious errors can occur if there is a regular spatial (e.g. topographic) pattern of variation in the grassland vegetation that could coincide with the grid. Such spatial patterns are generally obvious and appropriate sample location strategies can be used so that the results are not biased.

The sampling points in grid sampling are not independent and so standard errors of plot means cannot be calculated if the grid sampling occurs over a number of different plots. Standard parametric statistical analyses are therefore invalid and non-parametric methods should be used. However, in practice parametric analyses are often used and this should not have very serious consequences.

Stratified random sampling

Stratified random sampling is appropriate when the grassland area is clearly patterned. More detailed studies may then be required within the different communities so detected. Sample locations can be assigned to the different communities proportional to the area (strata) occupied by each. The locations assigned to each part of the pattern are then placed randomly within the different communities. More intensive sampling of particular communities may be desirable. For instance, sheep camps may occupy only a small proportion of a paddock but a higher sample density within the sheep camp and on the sheep camp margins than in the remainder of the paddock may be necessary to sample adequately the differences in botanical composition over small distances.

Size and shape of sampling units

Sampling units may differ in shape and size. In any type of grassland vegetation there may be large numbers of species present, differing in distribution, abundance and often sizes of the individuals. In addition, grasslands usually comprise both perennial and annual species and sometimes large numbers of one species of annual may appear at certain times of the year. It is obvious that if the individuals of a species are very small and there are large numbers of them, then a smaller sampling unit is necessary than for smaller numbers of larger individuals.

The distribution of individuals of the different species determines the shape of the sampling unit. If the individuals of a species are randomly distributed, then a square or circular sampling unit (quadrat) is satisfactory; but if the species of interest is highly clumped, then a rectangular quadrat will result in a lower coefficient of variation. If the entire species composition of the grassland is of interest, and if some species are randomly distributed while some are clumped and others are regularly dispersed, with a wide range of abundances, then a large rectangular quadrat would represent the best compromise.

A species area curve can be used to determine a quadrat size that will give adequate sampling of most of the species present in an area of grassland. The species area curve can be determined as follows:

1. Place a small quadrat (say 50 cm X 50 cm) on the ground and count the number of species enclosed.
2. Turn the quadrat over so that the sample area is doubled and count the additional species enclosed.
3. Do this repeatedly, keeping the area sampled as compact as possible, until no more species are recorded with further increases in the area.
4. Plot the number of species recorded against the total sample area. The point of inflection of the curve (marked with an arrow) where the number of species recorded starts to decrease as the area

increases is a compromise quadrate size for dealing with the majority of the species in the particular grassland sampled.

This method of determining a compromise quadrat size is not appropriate in highly patterned communities, because moving from one patch of the pattern to another will result in a rapid increase in the number of species recorded as the quadrat size increases.

4.4.1. Quadrate method

Quadrats

Simple quadrats of 50 cm X 50 cm or 1 m X 1 m are very commonly used for grassland studies but the size of the quadrats also depends on the vegetation type. In dense temperate grasslands, a commonly used size is 5 cm X 5 cm. These are easily constructed out of welded metal and are robust and useful in many situations. Quadrats made out of lOx 5 mm aluminium bar are light to carry and easy to see if mislaid. It is easier to make a square or rectangular quadrat to a precise size than it is to make a circular one. Rectangular quadrats should not be too long and thin as this increases errors associated with cutting or counting plants on the quadrat edge. A suitable size for a rectangular quadrat for many grassland situations is 1.0 X 0.5 m.

Closed quadrats are often difficult to place in very tall vegetation without flattening the plants and making it difficult to determine which plants are included within the quadrat and which are not. Three-sided or open-ended quadrats are much easier to push into the grassland close to the ground with minimum disturbance of the vegetation.

A more recent variation of simple quadrats is the use of nested quadrats. This arrangement overcomes the problems associated with requiring different quadrat sizes for sampling both rare and extremely common species within the one grassland community. Clusters of nested quadrats are set out in geometrically increasing areas (e.g. 1,2,4,8, 16 and 32 m^2) and species are recorded individually within these nested quadrats. In dense grassland with many species present, the size of the quadrats would be reduced. The use of nested quadrats is increasing because they are very efficient in terms of time invested relative to the large amount of information obtained, as discussed on pp. 89-91. A square shape is usually used and the corners of each sub-quadrat in the cluster are defined by marks on four diagonal cords leading outwards from a central peg.

Botanical composition expressed as proportion of individual plant species, is an important measure of vegetation, particularly in pasture studies. In routine, plant species composition is often estimated in the same quadrate in which herbage mass is assessed. The estimated composition is then used as weighing factor to improve the accuracy of pasture yield. This method works well if, there are few species of interest to evaluate.

Procedure

A quadrate of 1 m2 size is randomly placed at 6 locations in a pasture plot of I ha. The herbage inside the quadrate is cut 2cm above the ground level by shear and stored in a paper bag. Herbage sample is sorted in to different plant species. Individual species are weighed and dried in an oven at 70-80°C for 24hr. Then, the composition is expressed as proportion of different species in the sample on dry matter basis.

Advantages

1. The method is easy, when the number of plants species in the pasture are less.

Disadvantage

1. Requires more time in hand sorting, when a large number of species are present in the sample.
2. Quadrate cutting indicates overall change of herbage mass or individual species. It does not indicate how each species achieve that change, such changes may result from changes in the number of plant per unit area (plant density), the plant basal area or the mass per unit of basal area.

Further reading

Cobby J N, Ridout M S Bassett P J and Large R V.1985. An investigation in to the use of ranked set sampling on grass-clover swards. Grass and Forage Science. 40: 257-263.

4.4.2. 100 point frame method

This sampling device facilitates the recording of 100 point hits within a 30x60cm quadrate at a given sampling location. This technique provides quantitative data very sensitive to ground cover. The technique enhances in particular the ability to approach sample adequacy for minor species.

Design

The sampling device is made of two major components, the point frame and the quadrate base. The quadrate base of 30x60 cm is made from aluminum angle. At the end, 5 steel support tubes of 2.5 cm inside diameter and 6.3 cm long are installed. Similar number of tubes is also installed on other end. The tubes are placed 6 cm apart. Four legs, which supported the base, can be shortened or lengthened, depending upon vegetation height.

Frame shown in fig is 32-cm height and 65-cm width. Lower crossbar is positioned at 23 cm from that of the top frame. Twenty pins of steel welding rods, 91 cm long and 0.2 cm in diameter, sharpened to a needlepoint with a long bevel, are placed 3 cm apart in the frame.

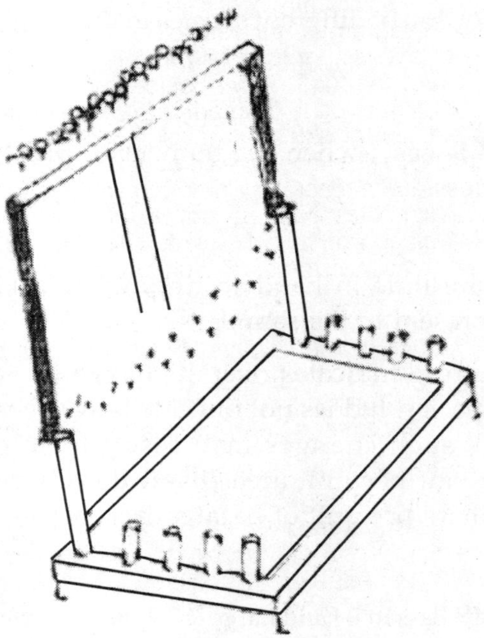

Fig. 9 : The point frame with quadrate base

Procedure

Once sampling location is identified, the point frame is placed in to the quadrate. Each pin is loaded separately and the pin hit is recorded for plant species at that location. After 20 pin hits are completed, the frame is removed to the next position on the quadrate base. When all 5 positions are observed, a total of 100 points is recorded.

Advantages

1. This technique provides quantitative estimates of the botanical composition
2. It is also helpful in estimating botanical composition of pasture's minor species.

Disadvantages

1. This method is unsuitable for large size plant species.
2. The method is impracticable for herbaceous species having large canopy cover

Farther reading

Faisal K Taha, Fisser H G. and Ries R E1983. A modified 100 point frame for vegetation inventory. Journal of Range Management. 36 : 124-125.

4.5. Transect Method

An extension of a rectangular quadrat is a transect which may be tens of metres or even kilometres long. A transect may be either in the form of a belt or a simple line. A belt transect usually consists of a tape or a string stretched across the vegetation, and species or individuals occurring within a specific distance perpendicular to the tape or string are recorded. A simple form of belt transect in large paddocks is to drive a vehicle across the vegetation and record species or individuals occurring between the tyre tracks. The vehicle must be driven along a predetermined course to avoid bias resulting from the driver avoiding difficult terrain.

With a line transect, individuals or species touching the tape or string are recorded. If changes in the sizes of the individual plants over time are to be followed, then the lengths of the intercepts occupied by individuals touching the line are recorded.

Transects are very useful in detecting and quantifying vegetation patterns within grasslands. For example, a 1000 m transect may be divided into 50 m segments with segments overlapping by, say, 20 m. Data would be collected from the 32 'segments' of the transect, each 50 m long, and then used for pattern analysis.

The transects can be laid out deliberately to intersect with patches of the grassland with different species composition and the edges of the patches can be precisely located. If these transects with the precise location of the edges are permanently marked, then changes in the size and/or species composition of the patches over time can be readily quantified.

This method holds good for estimation of botanical composition of grazing land where, the area is large and vegetation cover is diversely occupied by trees and shrubs.

Procedure Two 100x4 m transect are laid out at one site. Transects are marked/nailed at each end and a measuring tape is laid out between these points along the centerline. Every perennial plant whose butt fell partly or entirely within the confines of transect, as determined by a 1 meter measuring rod is identified and counted. Then, frequency of individual species from total number found in the sample is calculated.

Advantage

1. Useful where grazing lands is large and vegetation cover is diverse especially in shrubs and bushes occupied areas

Disadvantages

1. Minor herbaceous species are difficult to count

Further reading

Holm A McR, Curry P J and Wallace J F.1984. Observer difference in transect count cover estimates and plant size measurement on range monitoring site in an arid shrub land. Australian Rangeland Journal 6: 98-102.

4.6. Wheel Point Method

The size of a quadrat can be reduced to a point so that it has a location but no area. Points can be located randomly throughout a sample site, or systematically to ensure even coverage (at any scale but usually within representative parts of the area being sampled) or along a transect. Whether each point hits leaves, a plant base, litter or bare ground can be recorded. If 100 points are recorded and if a particular species is recorded as a hit in ten of these points, then it can be assumed that it occupies 10% of the area sampled. The recording of point quadrats can be rapid or slow, depending on how they are used.

Many different devices have been constructed to locate points within the area to be sampled. Some of these involve frames where the points are a fixed distance apart and the points are slowly lowered through the vegetation. Hits on leaves (aerial cover) as well as plant bases (basal cover) or bare ground can be recorded. The points in these frames are often fairly close together and if the individual plants in the grassland are large then one hit of a species can mean that the adjacent points hit the same plant of the same species and the observations are not independent. In addition, because grassland vegetation tends to occur in aggregates or clumps the use of a frame will lead to correlations within the points recorded each time the frame is locatedand the variance will be underestimated.

An approach to eliminate this autocorrelation is to distribute the points widely across the grassland being sampled. Methods to achieve this include the wheel point apparatus, where the spokes of a wheel are used to locate points regularly across the grassland. Usually one spoke of the wheel is marked and the data are recorded each time (or multiples of each time) the marked point hits the ground. Other methods of regularly locating points along a transect include cutting a notch in the front of the operator's boot, locating a small wire in the notch every predetermined number of paces, and recording the plant species that is hit.

An important source of error in point quadrat data collected by different operators lies in the definition of a 'hit', particularly when estimating basal cover. The point may descend within the perimeter of a living grass tussock, but if the point is very small it may only contact dead material. In addition, the tussocks of two separate species of grass may be interwoven and it can be extremely difficult to tell just which plant base has been contacted.

Theoretically points have location but no area and have investigated the relationship between area of the 'point' and the bias introduced into the results with respect to the estimation of cover. The larger the area of the 'point', the greater is the overestimation of cover. On the other hand, the smaller the area of the point, the more fragile the apparatus and the more difficult it is to use in the field. Sighting tubes with cross-hairs will reduce the effective area of the points but they are usually time consuming to use in the field. Caldwell *et al.* (1983) devised a fibre-optic point quadrat system with an effective point diameter of no more than 25 urn, The sensor is positioned at the end of a motor-driven rod that lowers the point through the canopy at a prescribed angle. When the sensor approaches a foliage element, the motor is stopped with the sensor within 1 mm of the plant part so that the operator can identify it and restart the system to proceed to the next contact. Some workers estimated this fibre-optic point system to be approximately twice as rapid as a conventional point quadrat system, even with motor-driven pins.

The precision of estimates of cover by species when using point quadrats is a function of the actual cover of a species and the number of points observed. For example, in a grassland with 10% cover and where one of the species present contributes only 1 % of that cover, there will be a 0.1 % probability of 'striking' the basal portion of that species, i.e. one chance in a thousand of striking the plant. Point quadrats are therefore not suited to sampling for rare species. In semiarid grasslands with low cover it is often recommended that 3000 points should be observed when monitoring cover by.

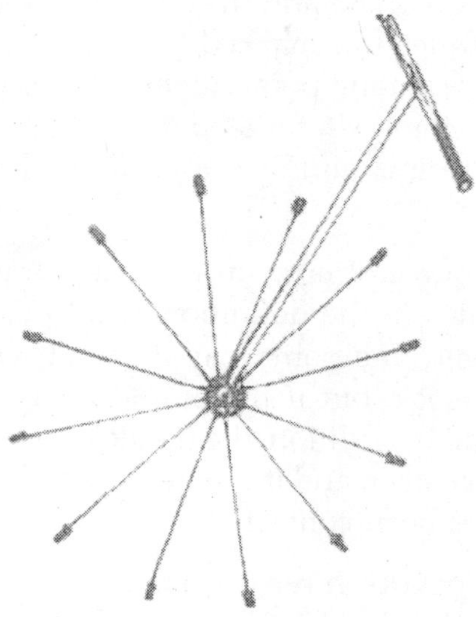

Fig. 10 : A wheel point apparatus

Procedure

An area of 50x50 m is marked in the pasture plot. By moving a wheel point apparatus, 1000 points are recorded. At each point, the individual plant species are identified. The frequency of individual plants species, ine:uding grasses, forbs and shrubs are expressed as number of total hit points.

Advantages

1. The method is simple and very easy to operate 2. Minor species can also be easily counted.

Disadvantages

1. In patchily distributed vegetation cover, this method alters the actual estimates.

Plotless sampling methods

Several methods have been devised using the distances from randomly distributed points to the nearest individual plant, or the nearest neighbour to randomly chosen individuals. These plotless methods can be very rapid and are particularly useful in sparse vegetation but are very sensitive to whether the species are randomly dispersed or not.

In the nearest-individual method, points are randomly selected and the distance to the nearest individual of a particular species is measured. It should also be noted that the relative abundance of species may be estimated by identifying the nearest plant to a randomly, or systematically, located point. The point to plant distance need not be measured.

In the point-centred quadrat method, points are randomly chosen and axes are set up dividing the space around the point into four quadrants. The distance to the nearest individual in each of the four quadrants is measured, which decreases the number of points chosen and the time in making the measurements. In sparse shrubland, range-finding devices have been used to measure the distances to the nearest individuals.

Cores

In very dense short grasslands, cores may be collected and taken to the laboratory for dissection and determination of the frequency or the rank order of dry weight of the constituent species. It also allows both above-ground and below-ground sampling of the same cores. Because of the large spatial variation in botanical composition, the location, size and number of cores is critical in terms of the statistical validity of the results. In general terms, a large number of small cores is more efficient than a small number oflarge cores. Core sampling is always destructive and nearly always involves laboratory analyses.

Fig. 11 : Recording the frequency of the species present using nested quadrats in a *Themeda* grassland. Four strings are laid out at right angles and the quadrat boundaries indicated using a tape between markers on the strings.The species present in the central quadrat are recorded first and then those present in the second quadrat, that is, the space between the two tapes. *The* tapes are successively moved outwards and the species recorded in each annulus until all quadrats in the nest have been recorded. Nests of seven to ten quadrats are usually used in grasslands. (Photograph C. Cooper.)

4.7. Composition by Mass

Methods of measuring the quantity of herbage mass or dry weight are described in Chapter 7. However, for many purposes it is necessary to partition the herbage mass into the individual species - that is, to measure the botanical composition by herbage mass. The most direct method is to harvest individual quadrats and physically sort into individual species or groups of species, dry and weigh them.

Clipping techniques

Clipping of quadrats followed by hand separation into individual species can be very time consuming if there are more than two or three important species in the samples or when a large number of samples are to be analysed. In addition, especially once they are cut, the recognition of single leaves of individual species (particularly grasses) becomes virtually impossible when there are many species. It may lead to a large percentage of the herbage mass being classified as unknown species. When separation into green leaf, stem and dead material as well as species is required the logistics can become virtually impossible. The accuracy of the estimates using clipping and sorting techniques is usually adequate, provided the number of samples taken is large enough, the botanical composition of the grassland is relatively simple and the material can be adequately identified. Another approach is to identify individual plants of the different species and cut them separately, putting the material of each species into a separate bag for drying and weighing. This procedure is time consuming, but does overcome the difficulty of identifying individual leaves of grasses in a mass of clipped material.

There are three major components of the herbage mass contributed by species. These are the number of plants per unit area (plant density), mean basal area of the individual species and the herbage mass per unit basal area. Information on all three components is desirable if the mechanisms of changes in herbage mass contribution of individual species in response to management are to be properly understood. For instance, addition of fertilizers may increase the mean basal area of an individual grass species but not increase the herbage mass per unit basal area. On the other hand, another species may respond in terms of increasing the herbage mass per unit basal area without much increase in mean basal area.

A technique for estimating the three components of herbage mass for individual grass species in complex native grasslands have been devised. Individual perennial grass plants were identified to species and clipped at ground level and the resultant herbage was later separated into green leaf, stem (including leaf bases) and dead material. The remainder of the herbage in each quadrate was also harvested. The basal area of each of the individually harvested plants was then estimated by placing a perspex sheet on the ground over each of the plants. The perimeter of each plant base was outlined and the outline transferred to a sheet of white paper, labelled, cut out with scissors and measured with an automatic leaf area meter. In some species, the contribution of herbage mass was more sensitive to plant density; basal area was important for the herbage mass estimates of all species tested and in some the mass per unit area was significant.

Estimation techniques

The use of visual estimates in the field is a very rapid procedure and allows the collection of data from a large number of quadrats. One of the early methods using estimates, and involving a long period of training, whereby operators estimated the herbage mass of individual tussocks with repeated checks of estimates against actual weights.

Three different methods for estimating composition are commonly used:

1. Dry-weight-rank

The DWR method is the simplest procedure for estimating the proportion by dry weight of the different species in a grassland. The method involves placing quadrats in grassland and estimating which species take first, second and third (or even fourth) rank in terms of dry weight. The species occupying rank 1 is given a value of 8.04, the species with rank 2, 2.41, and rank 3, 1.0. These multipliers have been derived empirically from studies of the relation between actual and ranked dry weight composition. However, their use has been successfully validated on a large number of grasslands comprising different species and different species composition. The above multipliers correspond to percentages of the components of 70, 21 and 9 for the first, second and third ranked species. The estimates are repeated on a number of quadrats and the values for each species are added up and expressed as a percentage of the total scores of all species.

DWR is surprisingly robust with respect to quadrat size. The major requirement is that the size be large enough for most quadrats to contain at least three species. The sparser the vegetation, the larger the quadrat size in order to contain the required number of species and quadrats up to 10 m^2 have been used. Used of 0.25 m^2 quadrat for estimating herbage mass and found that this was the largest quadrat that could be conveniently viewed as a whole in dense grassland (Figure 4.5). The choice of quadrat size rests with the observer and is a matter of compromise to obtain the most efficient use of time and the desired level of accuracy.

There are five important assumptions or requirements specific to DWR, the first four being discussed here. If the first two restrictions do not apply in any situation, these authors suggest appropriate modifications:

- At least three species should be present in the majority of the sample quadrats.
- If a particular species consistently has very high percentages (> 75%) in the quadrats, then a 'cumulative ranking' (first and second rank go to the dominant species) should be used.
- There should be variation among quadrats in the order in which species are ranked. If the order of ranking of the species in every quadrat is the same, then the results will indicate that the percentages of the species are 70, 21 and 9, respectively, in the area sampled if the standard multipliers are used.
- Species components, such as leaf and stem or green and dry, should not be treated as separate species as the standard multipliers are not appropriate.
- On the basis of computer modelling, DWR will not be appropriate if the species do not have some patchiness in their distribution. However, they concluded that only a small amount of patchiness is required and that in most grasslands this is probably fulfilled. This conclusion is substantiated by the successful validation of DWR on a wide range of grasslands.

Fig. 12 : Ranking the first three species in terms of their dry weight contribution (dry-weight-rank method) using a 0.25 m² quadrat and entering the results into a hand-held computer in a *Themeda* grassland. (Photograph C. Cooper.)

2. Direct estimation of percentage composition

The percentage contribution of each species to the herbage mass can be directly estimated for all of the species present in each quadrat rather than using the fixed multipliers of the DWR method. In other words, species A is recorded as being, say, 60%, species B as 25 % and so on. This procedure involves a higher level of precision than simply ranking species in order and it avoids the limitations of DWR listed above (Tothill *et al.*, 1992). However, it can be difficult to use in complex grasslands because it is not easy to estimate the percentages mentally so that they add up to 100%. It is also possible to estimate components such as percentage green leaf in the quadrat.

3. Percentage rank method

The percentage rank method combines the simplicity ofDWR with the greater accuracy of direct estimation. It is similar to DWR, in that only the first three or four species, in terms of dry weight, are 'rated', but rather than simply ranking them in order of weight, their actual percentage contribution is estimated. This also overcomes some of the limitations previously listed for DWR.

These two latter methods are of particular benefit in the following circumstances:

- One species is very dominant (> 90%).
- There are usually only two or three species in a quadrat with another minor species present in most quadrats, which would be overestimated byDWR.
- There are no dominant species, akin to the random distribution problem outlined by Neuteboom *et al. (1998).*

Estimating percentage directly, whether it be for all species or just for the three or four most important species, requires more training than simply ranking species as 1st, 2nd or 3rd.

Further reading

Wilson A D, Abraham N A, Barratt R. Choate J, Green D R, Harland R J, Oxley R E and Stanley R J.1987. Evaluation of methods for assessing vegetation changes in the semi-arid rangeland of southern Australia. Australian Rangeland Journal. 9 : 5-13.

4.8. Estimationof Botanical Composition of Diet

Knowledge of range herbivore diet is essential for efficient range management. This information is required for optimal forage allocation to different types of forage resources, selecting species for reseeding deteriorate ranges, predicting the outcome of overgrazing by different animals, identifying new species and determining the suitability of exotic animals for a particular range type. Presently much more information is needed if range management effectiveness is to be improved. The botanical composition of the grazing animal's diet is evaluated by

- direct observation of the animals
- utilization technique
- cannula techniques.
- stomach analysis
- faecal analysis

4.8.1 Visual observation of animals

Direct method

Procedure

An animal is continuously observed from a distance of 6-7 ft for collection of diet samples. The samples are collected for 45 minutes during active grazing hours. The samples are collected once daily for 2 days in the morning and 2 days in afternoon, so as to avoid any variation in actual sampling of diet composition due to diurnal changes in dietary preferences of animals. Actual portion of plant or plant parts are collected by simulating the animal's foraging behaviour. On an average forage samples from 30- 40 bites per hour are collected. After the collection period, individual species are sorted out and kept in an oven for dry matter estimation. Percent composition of individual species in the diet is then calculated.

Advantages

1. Accurate samples of the forage consumed by the animals are collected.

Disadvantages

1. Only one animal can be observed at a time for collection of the sample

2. Difficult to simulate and collect the actual lbrage Species and for parts consumed by the animals, in heterogeneous vegetation cover c f minor species
3. In small and dense ibliar cover, close observation is required. which disturbs the grazing of animals
4. Training of observer is mandatory
5. Animal needs to be tamed

Further reading

Reppert Jack N.1960. Forage preference and grazing habits of cattle at the eastern Colorado range station. Journal of Range Management. 13: 58-65.

112 Binocular method

Binocular Method

Procedure

The botanical composition of animal's diet on pasture is estimated by observing one animal carefully with a binocular (7x50 magnification) as it grazes. Samples of species that the animal grazed are collected on the same day, by hand clipping. Diet samples are sorted out after collection period, in to different species and dried in an oven. Then as percentage composition of each species in the diet of animals is expressed.

Advantages

1. This method has added advantages of not disturbing the natural grazing behaviour of the animal, as the animal is observed from a distance.
2. The method is simple and does not require sophisticated equipment.

Disadvantages

1. The method provides general qualitative information about plant species that are preferred by the animals, not quantitative one.
2. Only one animal can be observed at a time for collection of diet samples.
3. Difficult to differentiate between mere nibbling and active grazing.
4. Training of the observer is required.

Further reading

Galt H D, Theurer Brent, Ehrenreich H, Hale W H and Martin Clark .1969. Botanical composition of diet of steers grazing a desert grassland range. Journal of Range Management. 22: 14-19.

Microscopic technique

Hand clipped and oesophageal extrusa samples of diet and faeces are analyzed as follows

1. Hand clipped material

Procedure

Hand clipped sample is obtained by following the animals in the field. By simulating the animal's grazing pattern, forage samples are clipped and collected in paper bags. This material is frozen as soon as possible. It is later thoroughly mixed and sampled in the laboratory for percentage botanical composition in two ways.

Ten sub-samples are taken at random from the whole pile of material. Each of these is placed in a tray of approximately 5x30 inches size and spread evenly. The tray has a series of 25 notches on each side that serve as stops, when the tray is passed under a binocular microscope. One eyepiece of the scope has a cross hair. When the tray and material is passed under the scope, the species nearest to the cross hair is considered a hit. Fitly hits are to be recorded from each sub sample. Thus percentage botanical composition is determined by hits. It is expressed as number of hits in sub-samples.

$$\% \text{ of species} = \frac{\text{Number of individual species}}{\text{Total number of hits}}$$

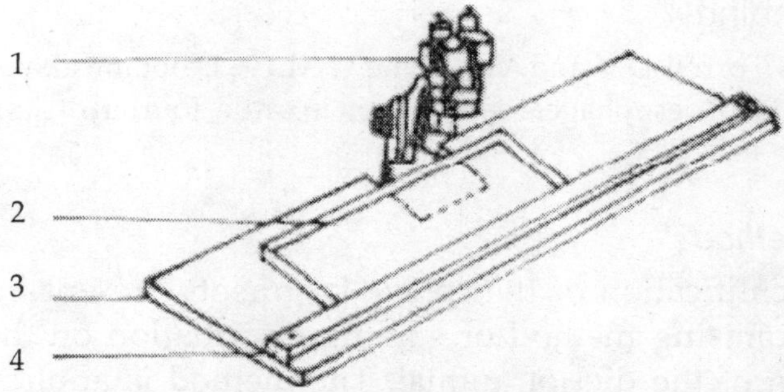

Fig. 13 : The binocular scope with accessories
1 Binocular scop 2 Tray 3 Base board 4 Movable guide

In the second method, each of these sub-samples are hand separated into different species, oven dried at 60-70°C for 24 hours, weighed and expressed as weight of individual species. Thus percentage botanical composition of clipped material is determined both by hits and by weight.

2. Oesophageal extrusa sample

Procedure

The extrusa sample is collected from a cannulated animal for 45 minutes during active grazing hours. The samples are collected once daily for 2 days in the morning and 2 days in afternoon. The sample is freezed, until further sampling is done. Samples are washed in cheesecloth and partially dried before the material is spread out on a slide for identification under a microscope. Eight sub samples with 50 hit each, are taken from each extrusa samples. Identification to each species and plant parts is recorded

The percent of plant species (X) is calculated by the following formula.

$$X = \frac{\text{Number of po int s on plant species}}{\text{Total number of po int s identified}}$$

Advantages

1. The method is more accurate compared to visual observation.

Disadvantages

1. The method requires sophisticated instrument
2. Trained person is required for the identification of plant species

Further reading

Harker K W, Torrell D T and Van Dyne G M.1964. Botanical examination of forage from oesophageal cannulation in cattle. Journal of Animal Science. 23: 465-469.

3. Faecal method

The identification of the cuticle fragments present in the faeces seems a promising method of gaining information on the botanical composition of the diet of animal. The method is applicable to any animal grazing any type of vegetation. The results are of most value, when considered in conjunction with a botanical analysis of the sward.

The basis of this method is that, the fragmentary material in the faeces contains fibres, hairs, vascular tissues and cuticle. Of these, cuticle bears sufficient characteristic features to be identifiable as belonging to a specific plant.

The cuticle is a waxy material scattered over the epidermis of the leaf When growth of tissue is complete, the cuticle is firm, impermeable to water and resistant to most chemical agents. The structure and arrangement of the epidermal cells appear to be characteristic of the species, thus the pattern distinguish on the cuticle is also characteristic.

Procedure

Preparation of reference slides

Fresh leaves are collected from available species growing in the pasture and from these a set of permanent reference slides is made. The leaves are macerated in 50% Nitric acid over a water bath and tissue washed in water to rerr.ove the acid. The cuticle is gently separated from rest of the materials using needle and a camel hair-brush. It is then floated on to a slide and cleared of any remaining cellular material as for as possible with a tine jet of water and a brush. The slide is taken through an alcohol series from 15 % to 95 %, stained with 1% Acid Fuchsin and permanently mounted in Euparal. Notes are made on the arrangement and size of the stomata and any hairs or glandular cell present by using microscope at 430 magnification. The permanent slides arc used as a set of reference for the identification of cuticic fragments found in the faeces.

Analysis of faeces samples

Eight fresh samples are collected from each locality, each sample is the dropping from one animal. The faecal samples are diluted with an equivalent volume of water and sub-sampled. A thin layer of material is spread between two slides and 50 identifications are made on each of four slides making a total of 200 identification per samples.

The action of digestion juices darkens the cellular tissues and imparts-a light brown stain to the transparent cuticle, making the cell outlines and other fractions distinguishable. Thus no special treatment or staining of the faecal material is necessary. Samples may be preserved with Formaline-acetic-alcohol if required and examined at a later date. In most cases it is possible to mark specific identification of the fragments under a magnification of 100 to 150.

Advantages

1. It does not interfere with normal grazing habits of animals as occur with use of cannulae and harness.
2. This method can be used in animals grazing on mixed communities, such as tropical rangeland

Disadvantages

1. An observer must have considerable training in order to accurately identify plant fragments.
2. Many plant species are difficult to separate at the species and some time at the genus level
3. Identification is further complicated by aging of faecal material before sample collection
4. Fragmentation may differ between species during digestion, so the relative proportion of species appears different
5. Extensive reference slides are required.
6. Cuticle of young plants readily disintegrates in the process of digestion thus can not be used for identification.

Further reading

Croker Barrara, H.1959. A method of estimating the botanical composition of diet of animal. New Zealand Journal of Agricultural Research 2:72-85.

Oesophageal cannula technique

Use of oesophageally cannulated animal is relatively successful in determining the diet of livestock species. It is extensively used in range herbivores species for accurate determination of diet composition.

Sample collection

Sampling period for collection of extrusa samples should be of 9 days (five days adaptation period, followed by 4 days for extrusa collection.). Animals are fitted with canvas collection bag with screen wire bottom. Oesophageally cannulated animals are grazed on the pasture. Extrusa samples are collected for 45 minutes during active grazing hours. The samples are collected once daily for 2 days in the morning and 2 days in afternoon. Extrusa samples arc mixed thoroughly by hand. subsequently pooled across the four days collection period for each animal.

Sample analysis by histological technique

Oesophageal extrusa samples arc oven dried and ground over a I min screen to reduce all plant fragments to a uniform size. Ground samples are washed over a 200- mesh screen to remove dirt and to remove very small plant fragments. A small portion of the mixture is spread evenly and mounted on a microscope slide using Hertwig's solutions and Hoycr's solutions. The slides are oven dried at 60°C. Five slides arc prepared from each sample of extrusa and examined under compound binocular microscope. Reference slide of plant tissues found in the sample is also prepared in a similar way for identification of species. Twenty locations are systematically observed on each slide. A location is considered as an area of the slide delineated by a microscope field using 125power magnification.

Identification of each, species in the mixtures is based on the epidermal characteristics:

- Forbs in the sample are identified on the basis of shape and size of epidermal hair and crystals in the epidermal cells.
- Grass species in the mixture are identified by occurrence and position of epidermal cells such as cork cells, silica cells, silico-suberose couples and aspirates.

Frequency percentage (number of fields that species occurred in out of 100 locations) is tabulated for each species in the sample. The frequency percentage are converted to particle density per field using a table developed by Fracker and Brischle (1944) and relative density expressed as a percentage of each species in the sample is calculated. The relative density of a species is used to estimate the percentage dry weight of that species in the sample.

Advantages

1. Used to identify forbs, grasses, shrubs and browse species generally masticated beyond recognition.

Disadvantages

- Relaiionship between relative density and relative percent composition by weight does not exist for all plant species.
- Sample preparation technique influences both accuracy and the ratio of the identifiable to non- identifiable fragments.

- Method gives poor estimation of certain browse species because, leaves from these species have a low proportion of epidermal material in relation to total biomass.

Further reading

Davies I. I 959. The use of epidermal characteristics for the identification of grasses in the leafy stage. British Grassland Society Journal. 14:7-16.

Fracker S. B and Brischle J A.1944. Measuring the local distribution of ribes. Ecology.25: 283-303.

Sparks Donnie R and Malecheck John C. 1968. Estimating percentage dry weight in diets using a microscopic technique. Journal of Range Management 21:264- 265.

Near infrared reflectance spectroscopy (N IRS)

NIRS has long been a useful tool for rapidly predicting nutritional characteristics of feedstuffs and forages. More recently, NI RS technology has been applied to forage, for estimation of plant part percentage. It has now shown potential for estimating botanical composition of complex mixture of diet and range.

Procedure

Oesophageal-cannulated animals arc grazed in the pasture plot. Extrusa samples are collected for 45 minutes during active grazing hours. The samples are collected once daily for 2 days in the morning and 2 days in afternoon. Extrusa is thoroughly mixed in the screened bottom collection bags, sub-sampled and later freeze dried. Dried samples are then ground to pass through a 1min screen.

Each sample mixture is packed in to sample cell with a near-infrared transparent quartz cover glass and scanned 64 times. Reflected energy (log I/R) is measured, averaged over 64 scans and recorded at 2 nm intervals from 1,100 to 2,500 nm. Two microcomputers based algorithms CENTER and SELECT are used to select, which served as reference samples. CENTER ranks samples in a file according to the H distance (Mahalanobis distance) from the average spectrum of the file. SELECT eliminates samples with similar spectra. Spectral data are transformed by a first derivative math treatment and principle components are calculated.

Some of the models available in the market are pacific scientific model 6250 scanning reflectance monochromometer interfaced to a Digital Equipment Company Micro-Vax 11 Minicomputer.

Advantages

1. Less time is required in analysis of samples.
2. It is easy to operate once the reference slides are prepared and standardized
3. NIRS offers accepted precision and accuracy in the prediction of major botanical components
4. It is practical and efficient, also reduces the number of samples required than other methods

Disadvantages

1. Salivary contamination of diet sample alters the spectral absorption and estimated values
2. Large number of reference slides of forage varying in age, species and season are required.

Further reading

VoleskyJerry D and Coleman Samuel W.1996. Estimation of botanical composition of oesophageal extrusa samples using near infrared reflectance spectroscopy. Journal of Range Management. 49: 163-166.

Laser induced fluorescence technique (LIF)

There is tremendous need for a fast, accurate and non-invasive procedure to evaluate botanical composition of free-ranging animal diets. Fluorescence is the re- emission of light by a material that has been exited by incident light; in LIF, the excitation source is emission from a laser. Typically the re-emitted light has a longer wavelength than the excitation light. The spectrum of fluorescence is characteristic of the fluorescing material and, hence, can serve for identification. A material's fluorescence can also depend upon the spectral character of the exciting light. All objects will fluoresce when exited with the proper wavelength.

Procedure

Oesophagcally cannulated animals are allowed to graze on a pasture. Extrusa and rectal grab samples are collected for 5 days. Faecal samples arc oven dried to a constant weight at 60-70°C. Individual pellets, each weighing approximately 0.32 g, are then placed in 160 vials each containing 20 ml of HPLC grade chloroform, capped and maintained at room temperature (19-24°C) until pellets sank to the bottom of the vial.

Individual pellets in 96 of the vials are crushed with forceps and three 4min diameter glass beads are placed in each vial. The contents are agitated on a vertex mixer until no particles larger than the glass beads are visible (2-3 mm). The pulvarised material is then poured into a 100m1 beaker and filtered through Whatman no. 4 filter paper. The 100m1 beaker is raised with an additional 5 ml of chloroform to remove all particles. The filtrate (extract) is re-captured in the original vials and the vials are recapped. Pellets are removed from the remaining 64 vials and recapped to prevent chloroform evaporation.

Six vials each of filtrate from the two diet components are also prepared for LIF measurements, in a beaker with 20 ml of chloroform swirling the mixture for 20s and then filtering as with the crushed faecal pellets. Another six vials containing only chloroform served as blanks. All 178 capped vials are placed in a freezer (-20 °C) until tested for LIF.

The 160 vials of chloroform exposed to faecal samples are tested in random order for LIF by pippeting 3m1 aliquots from each vial in to a 3.5ml quartz cuvette in a holder in the laser beam. Following exposure to the laser beam, the material in the cuvette is returned to its original vial, the cuvette is rinsed with pure 1-1PLC grade chloroform and the wash is discarded. The chloroform blanks, diet samples are evaluated at the beginning and end of the LIF testing and after every 30-33 faecal samples.

Each sample in the cuvette is excited by a series of four pulses of 355 nm light from a frequency-tripled neodymium yttrium aluminum garnet (Nd:YAG) laser. The pulses, spaced by 0.1 s, had duration of about Ions and energies of about 1 MJ. The resulting broad band fluorescence is input to a low-resolution spectrograph. Two long-pass colour filters are used at the spectrograph input to keep bright-scattered laser light from entering the instrument and possibly distorting the fluorescence spectrum. The dispersed spectrum at the spectrograph's out put is detected pulse-by-pulse using an intensified time-gated linear photodiode array. Output of the array, whose 774 central pixels covered the wavelength range from 400 to 800nm, is stored in a computer. The resulting spectra consisted of 'counts' (proportional to fluorescence light intensity) plotted against pixel number (corresponding to wavelength).

Calibration

The data analysis consists of sub-sample mean values of maximum counts in the blue and red and their corresponding wavelength. The mean (n=6 samples with four laser pulses per sample) spectrum of the chloroform only is considered to be the base line and is subtracted from each of the four laser pulses per sample (774 count -wavelength pairs produced per pulse), before selecting the maximum red and blue count and corresponding wavelength and red/blue count ratios, based on the sub- sample mean (n=4 laser pulses per sample) is to be analyzed.

Further reading

Anderson, D M, Nachman P. Este11 R E, Ruekgaur T, Havstad K M, Fredrickson E L, Murray L NV.1996. The potential of laser-induced fluorescence (LIF) spectra of animal faeces to determine diet botanical composition. Small Ruminant Research. 21:1-10.

4.9. Determination of Preference Index

Preference index (PI) is defined as per cent composition of species in the diet compared to per cent composition in the pasture. If PI of a particular plant species is 1, it indicates the preference for that species by animals is similar to their prevalence in the pasture vegetation. If it is <1. it reflects animal's avoidance and if it is >1, it reflects animal's preference for a particular plant Species.

Frequency measurement of plant species both in the diet and on the range is useful in interpreting forage preference because. the indices indicates not only the quantity of the plant the animals eat but also the consistency with which they select the plants distributed in the range.

Procedure

Oesophageal cannulated animals are used to collect diet sample (hand picked or mouth grab samples can also be used) for approximately for 45 minutes, two days in the morning and two days in the evening. The extrusa collected from each animal in a day is considered as one sample. The extrusa samples are analyzed by the microscopic point method under 15X magnification. Plant that could not be identified at this low power is mounted on a slide with Hoyer's solution and identified with a compound microscope at 125X magnification. One hundred points are examined for each sample to yield percentage diet composition. Frequency in the diet for each plant species is determined as percent of samples in which the species occurred. Plant species with in 6 quadrates of 1m2/ha are clipped and weighed to give an estimate of available herbage yield. The weight of each plant species present in each pasture quadrate is divided by the total weight of forage produce in the quadrate to yield percent range composition

Percent frequency on the range is computed for each species as the percentage of plot in which the species occurred.

For nutritional studies, the expression of PI on dry weight basis (dry weight of plant species in each quadrate over the total dry weight) is preferred compared to frequency basis.

Four ways for measurement of preference indices

(1) $$\mathrm{RPI} = \frac{\mathrm{fd} \times \mathrm{D}}{\mathrm{fr} \times \mathrm{R}}$$

(2) $$\mathrm{RPI} = \frac{\mathrm{D}}{\mathrm{fd} \times \mathrm{D}}$$

(3) $$\mathrm{RPI} = \frac{\mathrm{fd} \times \mathrm{D}}{100 \times \mathrm{R}}$$

(4) $$\mathrm{RPI} = \frac{100 \times \mathrm{D}}{\mathrm{fr} \times \mathrm{R}}$$

RPI = relative preference indices
fd = % frequency of plant species in diet
D = % diet composition
fr = % frequency of plant species on range
R = % range composition
fr x R
100 x R
ft x R

Disadvantage

1. In certain cases, the PI of a particular plant species may not be determined because either the plant is not sampled in the diet, while occurring in the pasture or sampled in the diet, but not in the pasture quadrate.

Further reading

Kruegar W C. 1972. Evaluating animal preference. Journal of Range Management 25:471-475

Van Dyne G M and Heady H F.1965. Botanical composition of animal and cattle diets on a mature annual range. Hil gradia. 36:465.

4.10. Diet Sampling Techniques

One of the basic problems confronting the range scientist is that of determining the nutritive value of the grazing animal's diet. Tropical rangeland are composed of diverse plant community and rugged terrain. An accurate and precise estimation of dry matter intake in animals on range are difficult to obtain. Information on the nutritive value of grazing animal's diet is essential for efficient livestock production.The diet samples also help in identification of the different plant species consumed by the animals, foraging strategy adopted by them to meet the nutrient requirement and conservation of palatable species. These methods are described below:

Clipping

A 1m^2 quadrate is randomly placed at 6 locations in a plot of 1 hectare. All the vegetation in a quadrate is cut with a sickle approximately 2 cm above the ground level and weighed. Then sub-sample is taken in a paper bag, kept in an oven at 60-70° C for 24h for drying.

Advantages

1. Requires little equipment and time to obtain a sample.

Disadvantages

1. It does not give a true representation of the diet sample
2. Preference indices can not be calculated by this method
3. This method does not consider the ability of the animal to select quality parts amongst the available herbage mass.

Hand Plucking

The animal is allowed to graze on the pasture. An operator in the pasture collects samples, by simulating the actual forage consumed by the animals. On an average 40- 50 bites per animal are collected for a period of 45 min for 2 days in the morning and 2 days in the evening. Samples are kept in paper bag and dried at 60-70°C for 24 h in an oven.

Advantages

1. This method is more reliable than the clipping method as it takes in to the actual dietary preference of the animal
2. The method is more suited for monoculture type of vegetation cover

Disadvantages

1. This method has limited use when plant occurs in intimate mixtures at a scale smaller than the bite
2. It has less reliability in places where plants are patchilydistributed.
3. This method is totally ineffective with complex mixture of forage
4. The method is difficult to adapt in ranges occupied by bushes and shrubs.

Mouth grabbing

The animal is allowed to graze on pasture, the forage consumed by the animal is immediately snatched before swallowing, by an operator. On an average 40-50 bites per animal are collected for a period of 45 min for 2 days in the morning and 2 days in the evening. The samples contaminated with excessive saliva are discarded. The samples are kept in paper bags and dried in an oven at 60- 70°C for 24 h.

Advantages

1. This method provides more accurate diet samples, compared to plucking and clipping.

Disadvantages

1. Frequent handling by the operator alters the grazing behaviour of the animal
2. There is possibility of sample contamination with saliva.
3. For frequent handling taming of animals are required
4. Samples are difficult to collect, when the bite size is very small.
5. Minor species are difficult to sample by this method.

Oesophageal extrusa method

Surgical procedure for fixation of oesophageal cannula

An animal is fasted for 24 hr before surgery and their mid ventral cervical region is shorn off and disinfected. One ml each of Sequil and Largactil is injected half an hour before surgery. Animal is laid on right lateral recumbancy on the surgical table. The animal is restrained properly and 3 to 4 ml of 2 per cent Xylocaine is carefully infused subcutaneously along the mid tracheal line in between the Omohyoideus muscle and trachea. Ten cm long skin incision is made along the traced line and subcutaneous fascia and connective tissues arc separated by blunt dissection. The incision is carefully extended between medial margin of Sternocephalicus muscle and lateral margin of Omohyoideus muscle. A rubber tube is passed through the oral cavity using a mouth-gag. The oesophagus is isolated by blunt dissection and exposed to view. A 5cm incision is made longitudinally through the oesophageal wall into the lumen. The flanged end of the polyethylene cannula is inserted in the lumen of the oesophagus. The anterior and posterior ends of the oesophagus are closed by inv-n-ting the edges of oesophageal submucosa with chromic cat gut suture. Subsequently purse string suture is applied to hold the submucosa of oesophagus around the tubular portion of the cannula and internal incision and submucosa is then closed by chromic cat-gut with continuous suture. The post operative care involved administration of parental antibiotic, analgesics and multivitamin preparations for 5 days followed by removal of skin suture and application of topical antibiotic and fly repellent.

Fig. 14 : In sin, oesophageal cannula

Fig. 15 : Cannukited animal with collection bag

Sampling method

Cannulated animals are fitted with canvas collection bags having, screened wire bottom for extrusa collection. They are allowed to graze freely in the pasture and samples are collected for a period of 45 min for 2 days in the morning and 2 days in the evening. The bags are removed after collection period and samples are thoroughly mixed by hand. Samples are dried at low temperature in an oven or freeze-dried in a lyophilizer.

Advantages

1. Choice of method for obtaining diet samples

Disadvantages

1. Requires surgery for fixation of cannula.
2. Difficult to collect samples for short period
3. Cannulation often alter the grazing behavior of animals
4. Ensalivation of the sample and high temperature drying increases the lignin value.
5. Ensalivation increases the nitrogen content of sample.
6. Chances of contamination of extrusa with regurgitated material.

Further reading

Bhatta R. Sankhyan S K Shinde A K and. Karim S A.1996. Nutritional evaluation using oesophageal ly cannulated sheep under silvi pasture. Indian Journal of Animal Sciences. 67:234-236

Karim SA, Sankhyan SK, Bhatta R and Shinde AK. 1997. Surgical procedure of oesophageal cannulation to assess grazing intake in sheep. Indian Journal of Small Ruminants. 3: 72-75.

Mc Innis ML, Vavra M and Kruegar WC. 1983. A comparison of four methods used to determine the diet of large herbivores. Journal of Range Management. 36:302-306.

Sankhyan SK, Shinde AK, Bhatta R and Karim SA. 1999 Comparison of diet and faecal collection methods for assessment of seasonal variation in dry matter intake by sheep maintained on a Cenchrus ciliaris pasture Animal Feed Science and Technology. 82. 261-269.

Torrell DT. 1954. An oesophageal cannula for animal nutrition studies. Journal of Animal Science. 13:878

4.11. Estimation of Feacal Out Put

In order to measure the amount of dry matter consumed by grazing animal, it is necessary to determine the indigestibility of the ingested herbage and the quantity of faeces voided per unit of time.

The faeces produced is estimated by two methods.

- By using faecal bags and harness
- By using indicators

Faecal bag method

Design

The faecal harness is comprised of 2 parts: A and B. Part A consists of 3 main bands namely neck, breast and lower abdomen band. On these, 3 bands of leather straps with buckles are attached. The part B of the faecal harness includes a rump band and the faecal bag. The rump band with its 4 straps can be buckled with the straps of the lower abdomen band. The faecal bag has hole in the bottom, to let urine flow (Fig. 16).

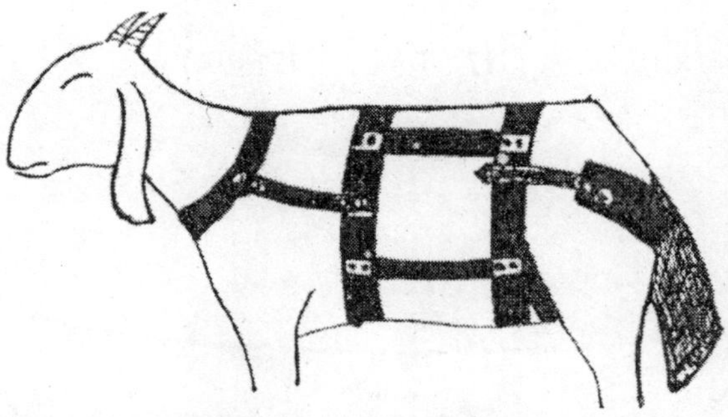

Fig. 16 : Animal with tie belts and faecal bags

Procedure

Animals are fitted with faecal harness at least 3 days before collection for adaptation followed by 6 days collection period. Faeces are collected every 12 hour interval or once in 24 hours depending on the volume of the faeces voided. The total faeces collected is thoroughly mixed. Sub samples are taken for the estimation of dry. matter and further analysis. The animal should be tied in faecal material free sheds to visualize any loses due to split during night hours. If any doubt existed as to whether faecal material has been lost, the data for the animal concerned are discarded for that day.

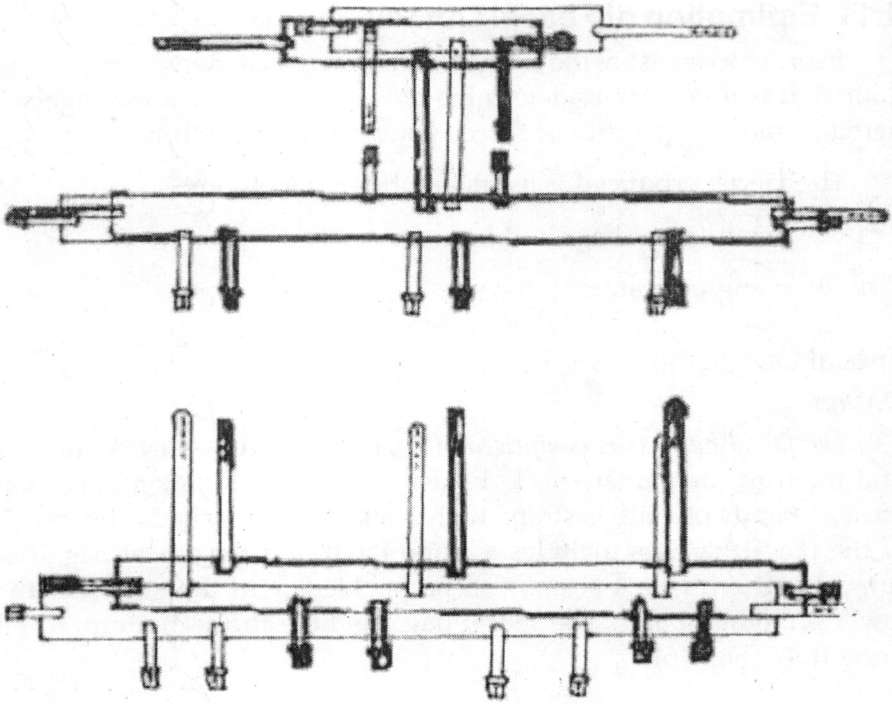

Fig. 17 : Tie belts (Part A)

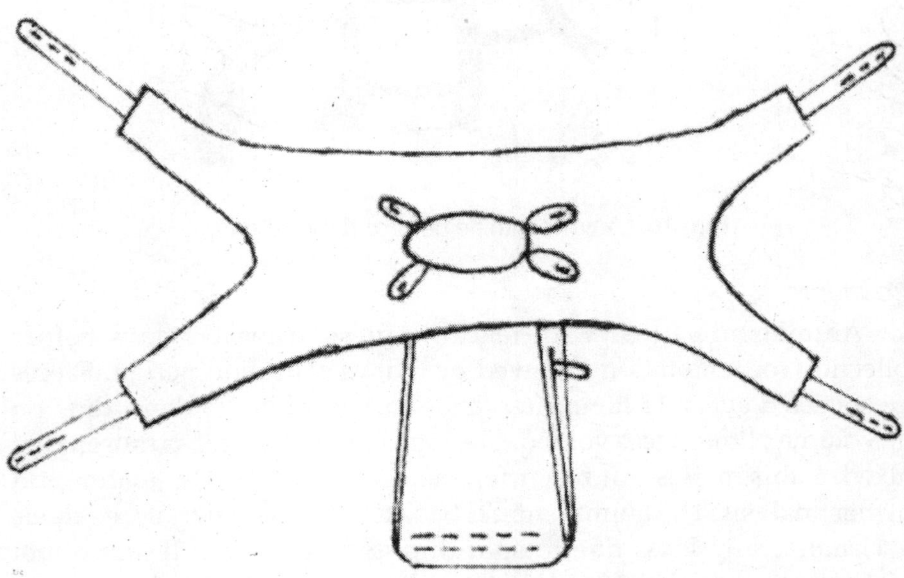

Fig. 17 : Faecal collection bag (Part B)

Advantages

1. Gives accurate estimates of faecal out go
2. No need to administer any marker/indicator to estimate faecal out go

Disadvantages

1. The animal must be adapted to the harness and frequent handling.
2. This method requires more labour for the collection of faeces
3. Harness is difficult to remove.
4. Harness is difficult to ensure in place particularly in shrub land.
5. Harness impairs the movement and bipedal stance of animal
6. It causes stress and discomfort to animal in some cases

Further reading

Doyle P T, Casson T, Cransbery L, Rowe J B.1994. Faecal output of grazing sheep measured by total collection or using chromic sesquioxide. Small Ruminant Research. 13: 234-236.

Yiakonlaki M D and Nastis A S. 1998. A modified faecal harness for grazing goats on Mediterranean shrub lands. Journal of Range Management. 51: 545-546.

Indicator method

Chromic oxide (Cr_20_3) paper capsule

Procedure

The grazing animal is dosed for 10 days with lg Chromic oxide (Cr_20_3) in a paper capsule twice daily at 8:00 and 17:00 h. The initial 5 days are allotted for uniform Cr_20_3 excretion and the later 5 days for collection of faeces samples. During the collection periods, rectal grab samples are taken both during morning and evening hours. Samples of faeces collected over 5 days periods are pooled and representative samples are dried in an oven at 60-70°C for 24 h.

Estimation of chromic oxide in faecal samples

One-gram oven dried faeces sample-containing approximately 10mg of Cr_20_3 is transferred to a 100m1- kjeldahl flask. 10 ml of concentrate nitric acid is added to it and kept for over night aging to make subsequent digestion easier. The mixture is heated over a micro burner to dryness on the next day. Add 10 ml of digestion mixture and continue the digestion until the green color changes to yellow, orange or red depending on the concentration of the chromic oxide. The flask is immediately cooled in the running water and make the volume to 100 ml with distilled water.

Preparation of digestion mixture

10 g of Sodium molybdate is dissolved in 150m1 of distilled water. Add 150m1 of concentrate Sulphuric acid, cool the solution in water bath. After cooling 200m1 of 70% Perchloric acid is added, raise the volume to 500m1 with distilled water.

Preparation of standard curve

10 mg Cr203 is weighed and transferred to kjeldahl flask. Similar digestion procedure as adopted for faeces samples is followed for standard preparation. The stock solution quantity ranging from 1 to 10m1 is transferred to test tubes and raise the volume to 10m1 in each test tube with distilled water so as to have concentration of Cr203 from 0 to 100 µg per ml. The optical density of these diluted samples is recorded at 430nm-wave length using spectrophotometer.

Stock solution (nil)	Distilled water (ml)	Cone. ot Cr_2O_3µg/ml	Transmittance (%T)	Factor (Cone./OD)
0	10	0		
1.0	9	10		
2.0	8	20		
3.0	7	30		
4.0	6	40		
5.0	5	50		
6.0	4	60		
7.0	3	70		
8.0	2	80		
9.0	1	90		
10	0	100		

Average factor (Conc./OD) – / ml

Total quantity of Cr_20_3 in the faecal sample is calculated from 100x factor x OD; if 2g of Cr_20_3 is administered to the animal and one gram of faecal sample is taken, then faecal out go is estimated as, 20,00,000pg/ (factor x OD of the sample).

Advantages

1. This method avoids the problems associated with the faecal bags

Disadvantages

1. Incomplete recovery of Cr_20_3 in faeces adversely affects faecal out go estimation.
2. Loss of Cr_20_3 due to regurgitation often bias the faecal out go estimates
3. Diurnal variation in Cr_20_3 excretion adversely affect faecal out go estimates
4. Re-ingestion of Cr203 by animals grazing on the same paddock alters the faecal out go estimates.
5. Sudden burst of capsule in the rumen results in uneven excretion of Cr_20_3 affecting the results.

Further reading

Hill F W and Anderson D L. 1958. Composition of metabolizable energy and production energy determination with growing chicks. Journal of Nutrition. 64: 587-590.

Krishna N, Bhandari D S and Patnayak B C. 1981. A double indicator method of determining dry matter intake and digestibility of nutrients in grazing sheep. Indian Journal of Animal Sciences. 51:716-719.

Sankhyan S K, Bhatta R and Karim S A. 1997. Relative efficiency of digestibility assessment in sheep by conventional and indicator method. Indian Journal of Animal Sciences. 67: 632-633.

Chromic oxide controlled release capsule (CRC)

Design

The capsule has a plastic barrel into which a matrix of Cr_20_3 sucrose mono-stearate (65:35w1w) is inserted. A plastic plunger and a compressed steel spring follow this. An orifice of variable size (depending on required release rate characteristics) expose the matrix to the rumen contents, forming a gel which is extruded through the force of the spring. A plastic strip, folded against the barrel for dosing opens into a T-configuration (wings) to retain the CRC in the rumen.

Procedure

CRC is administered orally to animal by a dosing gun, which releases the capsule at the back of the tongue. The swallowing motion of the animal then carries the capsules into the rumen. The wings of the CRC are temporarily restrained against the capsule barrel by a tape, which dissolves after 30-40 second in the presence of rumen fluids.

Seven days before the beginning of the collection periods animals are orally administered intra-ruminal bolus (Captec Chrome, Nufann Industries, Auckland, New Zealand) designed to release 201 mg of Chromic oxide/day. Daily sample of faeces is collected by grab method. The quantity of daily sample is about 30g, per animal and frozen for later laboratory analysis. Faecal Cr_20_3 concentration is determined by sulfuric/phosphoric acid digestion and bromate oxidation of ashed samples. After hydration of samples Cr203 concentration is measured using an atomic absorption spectrometer.

A mean Cr203 excretion rate across all animals is determined based on daily total faecal collection and fecal Cr,03 concentrations. These mean Cr_20_3 excretion rates are used in calculation to estimate faecal out go in place of manufacturer values.

$$\text{Faecal DM outgo}(g/d) = \frac{\text{Mean } C_{r2}O_3 \text{extraction rate, mg / day}}{\text{Mg of } C_{r2}O_3 \text{ / gm of fecal dry matter}}$$

Advantages

1. Single closing is sufficient
2. Reduces the disruption of animal's grazing behaviour by less handling
3. Excretion patterns of Cr203 is not influenced by the nature of diet consumed.
4. Reduces the labour requirement due to less handling of animals.
5. Samples of faeces can be taken at any time without fear of alteration in Cr203 content
6. Single grab sample per day is sufficient

Disadvantages

1. Chances of regurgitation and unexplained failure of bolus to release Cr_20_3 in the rumen may take place.

Further reading

Momont P A, Pruitt R J, Emerick R J and Pritchard.1994. Controlled release chromic oxide and alkane peroxide lignin marker methods. Journal of Range Management. 47: 413-423.

4.12. Estimation of Dry Matter Intake in Grazing Animals

The yield of animal products from grassland areas depends on a number of associated factors. To improve forage utilization, it is necessary to measure these components. One of the most important of these components is the quantity of forage consumed by grazing animals. Forage intake in grazing animal is estimated by

- Lignin ratio technique
- In-vitro dry matter digestibility technique
- n-alkanc technique

In all of these methods, accurate collection of diet and faecal samples is required. Four methods of diet sampling and two methods of faecal out go estimation in grazing animal have been suggested in previous chapter.

Lignin ratio technique

Collection of diet samples

The oesophageal ly-cannulated animal fitted with canvas sample collection bag having screen wire bottom are allowed to graze freely in the pasture. The extrusa samples are collected once daily for two days in the morning, and for two days in the afternoon for 45 minutes. After collection period, bags are removed and extrusa samples thoroughly mixed with hand and sub sampled for further analysis.

Other methods of diet sampling Viz., Plucking and clipping can also be used especially in monoculture type of pasture. In heterogeneous vegetation cover, oesophageal cannula technique is preferred method.

Collection of faecal samples

Animals are fitted with harness and faecal collection bags at least 4 days before collection for adaptation. Faeces are collected every 12 hours interval in the morning and evening for 5 days. In both the collection periods, faeces are collected, thoroughly mixed with the hand and sub– sampled (10% of total collection) for further analysis.

$C_{r2}O_3$ marker method in place of total collection by faecal bag may also be tried in situation where grazing land is occupied by shrubs and bushes and maintenance of faecal bags is difficult. The method for estimation of total faecal out go by $C_{r2}O_3$ has been detailed in previous chapter.

Analysis of lignia in diet and .faeces samples

Take 1g of dried sample and add 100 ml of acid detergent solution and reflux for an hour, filter in a Gooch crucible. Dry the crucible at 60°C for 8 hours and weigh. Fill the crucible containing ADF with 72% sulphuric acid and stir with glass rod to smooth paste and brake the lumps. Let glass rod remain in the crucible, refill with 72% H2SO4 and stir at hourly intervals as acid drains away. Crucible need not to be kept full at all times. Three additions would be sufficient. After three hours, filter off as much acid as possible. Wash the content with hot distilled water until it is free from acid. Rinse and remove glass rod. Dry the crucible at 60°C for 8 hours or over night and weigh. Keep the crucible in muffle furnace at 500°C for three hours, cool it and weigh.

$$\text{ADL}(\%) = \frac{(\text{Crucible after acid treatment}) - (\text{Crucible after ashing})}{\text{Weight of sample}} \times 100$$

Acid detergent solution: Prepare llitre of IN sulphuric acid solution, and add 20-g, cetyl tri-methyl ammonium bromide and mix thoroughly.

Dry matter intake of the animal is calculated by the following formula:

$$\text{DMI}(\text{kg.}) = \text{faecal outgo}(\text{kg.}) \frac{\text{lignin in faces}(\%)}{\text{lignin in feed}(\%)}$$

Advantages

1. Lignin ratio technique is the choice of methods when samples are carefully processed by freeze-drying.

Disadvantages

1. Lignin is digestible to some extent, thus this technique often under-estimates the dry matter intake
2. Formation of tannin-protein complex elevate the lignin value in tannin rich samples

3. Formation of artifact lignin during drying of samples at higher temperature elevate the lignin value
4. Freeze-drying of sample is costly and not readily available in most of the laboratories.

Further Reading

Crompton E W and Harris L E.1969. Applied Animal Nutrition. riedition. pp 120 WH. Freeman and Company. San Francisco.

Shinde A K, Sankhyan S K, Bhatta R and Karim S A. 1999. Comparison of lignin and 1VDMD techniques for estimation of DM1 and digestibility in range managed sheep. Indian Journal of Animal Sciences. 69: 336-338.

Van Soest P J and Robertson J B and Lewis B A. 1991. Methods for dietary fibre, neutral detergent fibre and non-starch polysaccharides in relation to animal nutrition. Journal of Dairy Science. 74: 3583-3597.

In vitro dry matter digestibility (IVDMD) technique

Currently in vitro digestibility of diet samples is the preferred method for determining dry matter intake in grazinc, animals. It is assumed that in vitro and in vivo digestibility of samples remain the same. However, in most of the cases it does not hold true.

Procedure

The collection of diet and faecal samples by various methods described in the previous chapters may be used.

Laboratory analysis

A representative sample of the diet is dried for 81) at 60°C in a forced draught oven and ground to pass through 0.8mm sieve. Rumen inoculum is collected from a cannulated animal maintained on hay based maintenance diet. The inoculum is strained through 2 layers of muslin (cheese) cloth in to a flask under anaerobic conditions by passing CO) gas, at 39°C. The 'synthetic saliva/ Mc Douglas's buffer' is prepared and CaCl2 is added after mixing all the contents. The solution is saturated with CO2 at 39°C. until the solution becomes clear.

Duplicate samples of oven dried forage (0.5 g) are weighed into a 100 ml flask, 40 ml of buffer solution is added, followed by 10 ml of strained rumen inoculum. The mixture is stirred and flushed with CO). The flask is then sealed with a rubber cork fitted with a Bunsen gas release valve. The 4-mm slit in the rubber tube on the valve is cut with a sharp knife. The slit normally remains closed. It is opened only to release the gas when the pressure inside the tube is more. After sealing the tube, the flask is incubated at 39°C in dark for 48h. The flask is shaken 3-4 times a day by hand. The residue is filtered off on tarred 12 cm No 4 filter paper or through porous alumina crucibles. The residue is dried at 60-70°C for 811 and weighed to determine the in vitro dry matter digestibility.

Calculation

$$\text{Dry matter intake}(\text{g day}) = \frac{\text{Facel outgo}(\text{g DM / day})}{1-\text{IVDMD}}$$

Advantages

1. Small quantity of sample is required.
2. Large number of samples can be analysed in short period.

Disadvantages

1. This method assumes that animal consumes a diet similar to that of animals for which faecal out put have been measured and that in vitro estimates of digestibility reflects in vivo digestibility. Both these assumptions do not hold true in all the cases.

Further reading

Tilley J M A and Terry R A.1963. A two-stage technique for the in vitro digestion of forage crops. Journal of British Grassland Society .18:104-111.

Van Dyne G M. 1969. Measuring the quantity and quality of the diet of large herbivores In: FG. Follet and HK Benchncr. A practical guide to the study of the productivity of large herbivores. Blackwell Scientific Publications. Oxford. pp 59-94.

n-alkane marker technique

Long chain fatty acids of chain length C19-C32, which are present in plant cuticular wax, have been suggested as internal markers as they are discrete compounds and are apparently indigestible. Suitable alkanes used as dosed markers are octacosane (C28) or dotriacontane (Cp) as they are easily obtained in pure form at low cost and present in herbage at very low concentrations.

Preparation of pellets containing C28-C32 alkalies

125-135 pellets from 13 sheets of Whatman No I filter paper are trimmed to fit a 35 x 49 cm aluminum tray. The tray with paper is heated in an oven at 100°C, 17g each of odd C28 and Cp are added. The sheets are separated and hung up to dry in a well ventilated room. The dried sheets are individually placed in an oven at 100°C for 2min to allow the paper to absorb the alkane. After cooling the paper is shredded with a guillotine. The weighed amount of shredded paper for each pellet is compacted by squeezing through a tube of 1-2 cm sheet of tissue paper using starch paste as adhesive. The papers arc dried at 40°C.

Dosing of n-alkanes

Each animal is administered alkane pellets by a dosing gun before allowed for grazing in the morning for a period of 16 days.

Diet sampling

The oesophageal ly-cannulated animals fitted with canvas sample collection bag with screen wire bottom are allowed to graze freely in the pasture. The extrusa samples are collected once daily for two days in the morning and for two days in the afternoon for 45 to 60 minutes. After collection period, bags are removed and extrusa samples thoroughly mixed with hand and sub sampled for further analysis.

Faecal sampling

Five days after beginning of the .alkane dosing, daily faeces collection are carried out for 11 days; on the final 8 days sub-samples of the daily faecal output are taken and bulked for each animal. For 24h after completion of the balance period, faeces are collected for every 3h from each animal and dried in oven at 60-70°C.

Analysis of alkane

Samples of freeze dried herbage (3g) and concentrate (3g) are extracted for 6h with petroleum spirit (B.P. 40-60°C) in a soxhlet extraction system with prior addition to the extraction thimbles of 0.6mg tetracontane (C34, Sigma London Ltd,Poole, Dorset) as an internal standard. Faeces sample (1g) are extracted, together with 0.6mg C34 internal standard, with the same solvent for 40 min in a Soxhtec system. The lipid extracts are transferred to I 00x2Omm thick walled screw-topped pyrex tubes. After removal of residual solvent

5inl IM alcoholic KOH solution is added and tubes are stoppered with PTFE-lined caps. The tubes are heated over night in a dry-block heater at 90°C. After partial cooling 5m1 hexane and 2m1 water are added and tubes are shaken vigorously. The top (non aqueous) liquid layer is removed, evaporated to dryness, redissolved in 1.5ml hexane and applied to top of a small column (disposable polypropylene column, Supelco Inc.) containing silica gel (Kieselgel 60, 70-230 mesh, Merck) with a bed volume of 5m1. The hydrocarbons are eluted with 10m1 hexane. The elute is concentrated to 150-300d then injected on to a 1.6mx4mm glass column containing 3% SE-30 on 100-120 mesh Supelcoport in a Model 104 gas chromato2raph fitted with a flame ionization detector. The chromatograph oven is maintained at 275°C and carrier gas, N2, with a flow rate of 30m1/min. Peak areas of the n-alkanes are determined using a Pye Unicam CDPI computing integrator. The carbon chain lengths of the nalkane arc deduced by their retention times relative to known alkalies.

The alkane contents of the dosed pellets are determined by extraction of complete pellets for 10h in a Soxhlet extractor using petroleum spirit (B.P.40-60) after weighing 1 20mg, C24 internal standard into the thimble and perforating the paper wrapper of the pellet. A portion of the extract is then submitted to the same analytical procedures as the dietary and faecal extracts.

The herbage intake is calculated from a pair of a natural dietary alkane, i and j as follows:

$$\text{Herbage intake}(\text{kg}) = \frac{\text{Fi}/\text{Fj}(\text{Dj}+\text{Ic.Cj})-\text{Ic.Ci}}{\text{Hi}-\text{Fi}/\text{Fj}-\text{Hj}}$$

where, Hi, Ci and Fi are the respective concentrations (mg/kg DM) of the natural odd chain alkane in herbage, concentrate and faeces: Hj,Cj and Fj are the respective concentration (mg/kg DM) of the even chain alkane in herbage, concentrate and faeces; lc is the intake of concentrate (kg DM/day) and Dj is the amount of alkane J. dosed by pellet (mg/day)

The intake of herbage is also calculated from the C35 contents of herbage, concentrate and faces and from the faecal out put using following equation:

$$\text{Herbage intake}(\text{kg DM}/\text{day}) = \frac{F_0.F_i - I_c.C_i}{\text{Hi}}$$

Where, Fo represents faecal DM out put (kg/day).

Advantages

1. Herbage intake estimation from long chain thtty acids is not affected by feeding level or presence of concentrate in the ration.
2. Saliva content of extrusa samples has little effect on alkane content.

Disadvantages

1. All the forage species found in the study area is to be analysed lbr long chain fatty acids (C12-C32) content.
2. Long chain n-alkanes are absorbed to a certain extent in monogastric animals.
3. Requires at least 6 days of dosing with alkanes before commencing faecal sampling.

Further Reading

Dove H and Meyes R W. 1991. The use of plant mix alkanes as marker substance in studies of the nutrition of herbivores. A review. Australian Journal of Agricultural Research. 42: 913-952.

Estimation procedure for digestibility estimates are described in vivo chapter

4.13. Measurement of Energy Expenditure in Grazing Animals

Knowledge of whole body energy expenditure is important if we are to make accurate estimates of energy requirements for optimum health, performance and production in animals. Energy requirement varies with nutritional, genetic and environment influence and these can interact in complex ways, therefore, it is necessary to isolate and identify specific factor which are important in controlling energy expenditure. In order to do this, it is useful to reduce whole body energy expenditure to smaller component parts; e.g. basal metabolic rate (BMR), dietary induced thennogenesis, the energy cost of activity, growth and lactation. In highly controlled and defined conditions like in calorimetry chamber, it is possible to investigate the effect of nutritional, genetic and environmental influence on each of the components of energy expenditure separately. However, such measurements may not be representative of the true energy expenditure (and hence energy requirements) of free living animals. In order to measure these, we need a technique, which will allow the measurement of energy expenditure without restriction.

Alternately, energy expenditure can be estimated from calorimetric determination of the energy cost of various activities and using estimates of the time spent in each of these activities. A value of total free energy expenditure can be arrived at. In addition to these approaches, there are two isotope techniques, which measure carbon dioxide production isotopically,

- labeled bicarbonate method
- doubly labeled water (DLW) method

CO_2 entry rate technique (CERT)

The CO_2 entry rate technique (CERT) is developed to estimate the energy expenditure of grazing animal by infusion of (14_c) – bicarbonate and calculation of dilution of (14_c) – bicarbonate, by the unlabelled bicarbonate pool.

Procedure

Animal is prepared for infusion and sampling I day bethre the start of the $NaH_{14}CO_3$ infusion. At that time, an intra abdominal catheter is placed in the animal's right side into the peritoneal cavity. Sample collection catheters of polyethylene tubing with perforation near the tip are placed into the parotid gland ducts. Tubing exited the mouth via a perforation of the cheek, is attached by elastic tape and glue to the top of the animal's head passed down the midline of the neck and is connected to a portable peristaltic pump housed in nylon back packed on the animal's left side. The infusate is kept in a collapsible plastic infusion bag infused at a rate of 4.6 ml per hour by another portable peristaltic infusion pump housed in a second nylon back pack on the animal's right side. The saliva sample is collected in to a glass bottle at a rate of 4.6ml per hour. The concentration of CO_2 is determined by placing 15ml of saliva in a 250m1 Erlenmeyer flask. A glass scintillation vial containing 3m1 of carbonate free IN NaOH is suspended inside the flask to trap the evolved CO_2. The sample then is acidified with 6m1 of IN H_2SO_4 injected through the rubber stopper. After 2hr, the vial is removed and 1.5ml of IM $Bac1_2$ is added to the vial to form insoluble $BaCO_3$. CO_2 concentration is determined by titrating the excess NaOH with 0.18N HC1. The specific activity of the ^{14}C is determined after adding 10m1 of scintillation cocktail shaken to homogenize the suspension and counted for 10min. CO_2 production is calculated as dilution of the infused ^{14}C activity by non-radioactive carbon in the bicarbonate pool. Energy expenditure (Y) is calculated from the equation of the Engels et al. (1976).

$$Y\ (kJ.day^{-1}) = 17.55\ LCO_2 day^{-1}\ -890$$

Disadvantages

1. The animal must carry pumps to infuse the isotopes and to obtain blood, urine or saliva samples
2. Experimental animals must be tamed.
3. Overestimates CO_2 expiration. if the infusion sampling protocol lasts less than 20h.

Further reading

Eagels E A N, Inskip M W and Corbett J L. 1976. Effect of change in respiratory carbon dioxide entry rate in sheep and their energy expenditure. Publication En. Association Animal Production. I 9: 339-342.

Whitelaw E G, Brockway J W and Reid R S. 1972. Measurement of carbon dioxide production in sheep by isotopes dilution. Journal of Experimental Physiology. 57:37-55.

Double labeled water technique

This method measures energy expenditure ot animals from the differences in turnover rate of the hydrogen and oxygen of body water. When body water is labeled with D2180, the difference between the two turnover rates can be measured and employed to calculate the rate of $_{CO2}$ production (r $_{CO2}$). In its simplest form r $_{CO2}$ is calculated on the 180 flux minus the deuterium flux times 0.5. The equilibrium fractionation factors for deuterium (f1) and 'NO (f2) between water vapour and water liquid at 37°C is 0.941 and 0.9925, respectively. However there is another fractionation factor relevant to the calculation of r_{CO2} and that is due to the fractionation which occurs when heavy water —0 equilibrates with CO, -0 (f3). Assumint; that fractionation is an equilibrium process these factors can be used to calculate the water flux (rw) and hence r CO,. If the proportion of water which is lost evaporating (x) is known,

$$rwD\ [(\ fi.X+ (1-X)]\ -1$$

$$rCO_2 = \frac{0 - (f2.x.rw) + (1 +x)\ rw}{2.f3}$$

The D and O flux rates are calculated from the rate constant of isotope decay times pool size. These can be computed from exponential curves fitted to multiple isotopes enrichment data points, from the gradient of a straight line fitted to log- transformed multiple-isotopes enrichment data points or by fitting a straight line between two log-transformed data point, at the beginning and at the end of the study. Energy expenditure is then calculated by the equation of Weir (1949).

EE (kJ) = 16.50 x RQ (I t.)+4.63 x rCO, (lit.) - 9.08x nitrogen (g)

Advantages

1. It gives integrated values for r_{CO2} over longer periods (typically 1-3 weeks).

Disadvantages

1. It can not be used for short periods
2. It is less precise than respiration chamber measurements.
3. It only gives a value for rCO_2. therefore it is necessary to know the RQ to calculate heat production.
4. It is more expensive than chamber studies
5. Cost of 180 is very high
6. The technique has little application in pastoral research.

Further reading

Lifsou N, Gordon G B and Mc Clintock R. 1955. Measurement of total carbon dioxide production by means of D20 . Journal of Applied Physiology. 7: 704-710.

Weir JB DE V.1949. New methods for calculating metabolic rate with special reference to protein metabolism. Journal of Physiology. 109:1-9.

Tracheal cannula technique

Surgical procedure for fixation of tracheal cannula

The animal is fasted for 24 hr before the surgery and mid ventral cervical region is shorn off and disinfected. One ml of tritlupromazine hydrochloride (Sequil) and chlorpromazine hydrochloride (Largactil) are injected half an hour before the surgery. The animal is placed on the operation table on dorsal recumbency, restrained properly and 5 to 6 ml of 2 per cent lignocaine hydrochloride (Xylocaine) is carefully infused subcutaneously along the traced mid ventral line of the trachea.

Five cm long skin incision is made along the traced line and subcutaneous fascia and connective tissue are separated by blunt dissection. Two bands of Omohyoideus muscle are separated by blunt dissection using alley forceps. The incision is carefully extended, exposing about 6 tracheal rings. A mid ventral incision is made cutting the 2 annular rings of the trachea and the cannula is fixed with its long axis facing forwards the nasopharynx. The cannula is secured in place by applying purse string suture using chromic cat-gut no.2-0. Internal tissues are closed by absorbable continuous suture and skin incision is closed by interrupted suture using silk thread. Postoperative care involved administration of parentral antibiotics, analgesics and multivitamin preparations.

Collection of expired air

An inflatable balloon is introduced into cannula so that. inspired and expired air could be diverted through the cannula using latex valve. The expiratory side of this valve is then connected by short length (8-) of non- collapsible tubino, to a meteorological balloon. The expired air is collected in the balloon of 25-30 litre capacity for 5-10 min in each recordino, period until the balloon is completely tilled but not distended. Volume of the air in balloon is determined by the formulae,

Volume (litres) = 4/3 71 r 3

Where, r = radius of the inflated balloon (cm)

The volume of the expired air is then corrected for standard temperature and pressure (STP) by taking the temperature and pressure of air in the balloon. 02 and CO_2 content of the expired air are determined using haldane gas analyzer.

Fig. 18 : *In situ* oesophageal cannula

Fig. 19 : Cannulated animal with collection bag

Analysis of expired air

Haldane gas apparatus

10 ml of expired air is drawn from balloon into a burette and mixed with CO, absorber (20% KOH solution). The change in level of mercury between initial and after absorption of CO, indicated the amount of CO, present in the expired air. Similarly the 02 is absorbed in pyrogallol solution (10g pyrogallol dissolved in 100 ml of saturated solution of KOH) and change in mercury level after absorption of 02 indicated the amount of 02 present in the expired air. The per cent composition of $_{CO2}$ and 02 in expired air is calculated. Energy expenditure (EE) is calculated according to formulae of Young and Webster (1963) excluding the CO, and .CH4 factors. Thus

EE (kJ) = 20.28 x V 02 (litre at STP)

Where, V02 is volume of 02 consumed by the animal

Advantages

1. The contamination of expired air with belching gases can be avoided.

Disadvantages

1. Fixation of tracheal cannula is cumbersome
2. Survivability of animal over a long period is a serious consideration
3. Restriction of air flow through nasopharynx, grossly disturbs the process of olfaction and thermo-regulation
4. Stenosis of tracheal rings alters the respiration rate

Further reading

Karim S A. Bhatta R, Sankhyan S K and Shinde A K. 1997. Surgical procedure of tracheal cannulation in sheep. Indian Journal of Small Ruminants. 3: 34-38.

Shinde A K. Karim SA, Sankhyan S K and Bhatta R.1998. Seasonal changes in physiological responses and energy expenditure of sheep maintained on semiarid pasture. Journal . of Agricultural Science, Cambridge. 131: 341-346.

Young B A and Webster M E D. 1963. A technique for estimation of energy expenditure in sheep. Australian Journal of Agricultural Research. 14: 867-873.

Facemask technique

The measurement of energy expenditure by tracheal cannula is cumbersome and also difficult to maintain the tracheal-cannulated animal for longer periods. Use of facemask for measurement of gaseous exchange and energy expenditure is an ideal method.

Measurement principles

The respiratory equipment consists of animal facemask and meteorological balloon. It works on the principle of gaseous exchange measurement. The energy expenditure is calculated from recording of CO, production and 02 consumption.

Facemask

The mask is made up of transparent PVC sheets and attached to animal by means of an adopter. The air tight end of the mask is fixed to the animal's face with an adapter. The air stream is separated by means of latex valve of various sizes. Two lateral inlet valves (10mm) are fixed in the mask at nostril level, while one outlet valve (30mm) is fixed in the middle. This arrangement ensured free movement of air without any distress on the animals.

Collection of expired air

The outlet valve is connected to a meteorological balloon by a 1 m long PVC tube (diameter 20mm). The expired air is collected in a balloon of 25-30 litre capacity for 5-10 min in each recording period until the balloon is completely filled but not distended. The volume of air collected in the balloon per unit time is determined and corrected for standard temperature and pressure. The 02 content of the expired air is determined by gas analyser.

The procedure for 02 estimation in expired air is described in the earlier section.

Energy expenditure is calculated according to the formula of Young and Webster (1963) excluding the CO, and CH4 factors. Thus

Energy Expenditure (kJ) = 20.28 x V 02 (litre at STP)

Where, VO, is volume of 02 consumed by the animal

Fig. 20 : Face mask in situ

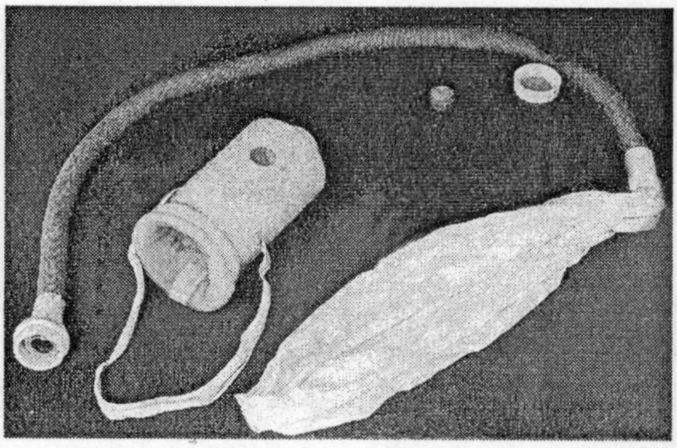

Fig. 21 : Face mask prototype

Advantages

1. Facemask is simple and handy.
2. Expired air can be collected at any place and any number of times without distress on the animals
3. Face mask is superior to tracheal cannula as a routine method of collecting expired air from animals

Disadvantages

1. Possibility of contamination of expired air with belching gases
2. Face mask prevents grazing while collecting the expired air
3. In some cases fixation of mask increases respiratory rate of animal
4. Tamed animals are required

Heart rate correlation method

Measurement of energy expenditure from gaseous exchange of an animal is technically difficult and not generally feasible for animal grazing under field condition. It is well known that a linear relationship between oxygen consumption and heart rate exists. This relationship could be used to calculate the energy expenditure of animal from heart rate.

Procedure

The heart rate of animal at regular interval is recorded by stethoscope placed on the ventro-lateral thorax. In the recent years, electro-cardiogram cassette system is used for accurate measurement of heart rate in animals, under field conditions. The ECG recording cassette is held in place over the scapula. One of the electrodes is placed on the animal over the first rib behind the scapula about half way down the body and the other over the scapula itself. Shaving off the hair in the selected site, washing the skin with alcohol and applying a conducting jelly to the electrode increases electrical contact. Heart rate is measured using a specially calibrated meter, which integrates the time taken between successive QRS complexes on the printout paper to give beats per minute. The energy expenditure is calculated by using the heart rate value in the formula

$$Y = 0.13 + 0.50\ X$$

When animal is at rest

$$Y = 2.86 \pm 0.88\ X \ [2]$$

When animal is grazing in pasture,which includes cost of walking and out door activities. Where, Y = energy expenditure (Kcal/hr)

X = Heart rate (beats/min)

Advantages

1. Simple and easiest method once the regression equation between energy expenditure and heart rate is established.

Disadvantages

1. Relationship between energy expenditure and heart rate varies with age, season, period of the day during which the measurement is recorded.

Further reading

Shinde A K, Karim S A. Singh V P and Sankhyan S K. 1999. Use of heart rate to predict energy expenditure of sheep on rangeland. Indian Journal of Animal Sciences. 69: 321-323.

□□□

Chapter 5

Analysis of Chemical Constituents in Feed and Plant Materials

5.1. Proximate Analysis

The proximate analysis is a scheme for routine description of animal feedstuffs devised in 1865 by Henneberg and Stohmann of the Weende Experiment Station in Germany (Lloyd et al., 1978). It is often referred to as the Weende System and was principally devised to separate carbohydrates into two broad classifications: crude fiber and nitrogen free extract (NFE). The system consists of determinations of water (moisture), ash, crude fat (ether extract), crude protein, and crude fiber. As indicated, NFE is a component of the system, but it is measured by difference rather than by analysis. Students should be aware that the proximate analysis system is both comparative and predictive in nature, as this will aid in understanding the broad purposes for development of the system.

First, let us consider the comparative aspects of the system. Most people, even those unfamiliar with livestock feeds and feeding, could look at a sample of first-cut alfalfa hay and a sample of dormant winter range grass and readily surmise that the alfalfa was a higher quality feed than the range grass. Alfalfa simply looks better, or at least from the human vantage point, it has more eye appeal than dormant winter range grass. Fewer people, however, could tell how much better the alfalfa is in terms of any specific nutrient (e.g., protein, fiber) or the production either feed will support. Thus, one important reason for development of the proximate analysis scheme was to allow comparison of feeds on a specific basis. It is often stated that one can not compare apples with oranges, but one can compare the protein as a percentage of dry weight in apples and oranges, and in doing so, make some realistic judgements about the nutritional value of each fruit. By the same token, proximate analysis allows one to make legitimate comparisons of feeds on the basis of specific nutrients, allowing one to judge how much better one feed is than another in terms of specific nutrients.

Now, what about the milk production or growth rate alfalfa hay will support compared with the dormant range grass? This question is addressed by the predictive nature of proximate analysis. If we know all the proximate components of various feeds, and the production the feeds will support, we might be able to develop regression equations to help us predict performance by livestock fed these feeds. In the more typical case, we try to use proximate components to predict factors related to performance, like digestibility and intake and then relate these predicted values to estimates of performance. As will be noted later, the proximate system has some failings that prevent it from being an extremely valuable predictive aid, and considerable research has been conducted in recent years to refine and add to the basic system to make it a better predictive tool.

To summarize, the proximate analysis system is an old scheme of laboratory analyses that allows comparison of feeds on the basis of specific nutrients and, to some extent, prediction of components of animal performance. Next, we will take a look at the specific analyses involved in the proximate analysis scheme.

5.1.1. Dry matter and ash

Dry matter or, more specifically, moisture determination is probably the most frequently performed analysis in the nutrition laboratory. It is an important analysis, in that the concentration of other nutrients is usually expressed on a dry matter basis (as a percentage of the dry matter). Because most students seem to have an inordinate amount of difficulty with conversion of nutrients to a dry matter basis, an example showing how this is done, as well as an example of dry matter determination calculations, is given in Table 3-2.

Moisture or dry matter content is extremely important to the livestock industry, particularly those segments that deal with high-moisture feeds. Consider, for example, a feedlot with 20,000 tons of silage inventory. Obviously, as silage is removed from storage, its dry matter content can change, resulting in a change in the feedlot's inventory of silage dry matter. Without routine dry matter determinations, the feedlot could grossly over or underestimate its inventory and perhaps over or undercharge customers for feed. There are numerous examples of the importance of accurate measurements of moisture in the livestock industry that should be readily apparent to most students.

Students typically believe that dry matter analysis is the simplest analysis they perform in the lab. For the most part, this is true because it routinely consists of weighing the sample into a tared (previously weighed) pan, placing the sample in a 100 to 105°C oven for 12 to 24 h and reweighing. Moisture content is simply the loss in weight from evaporation of water. This procedure works well for most feeds, but with some feeds, especially high-moisture, fermented feeds, some problems can be encountered. High-moisture feeds usually contain volatile nutrients that can be lost with 100°C oven drying. Volatile nutrients of greatest importance are short chain fatty acids (acetic, propionic, butyric, etc.), but essential oils (menthol, camphor) also can be important with some feeds. Drying samples at 100°C can volatilize some of these materials, resulting in greater moisture (lower dry matter) values than expected. This concept is considered in some detail along with some other important points in an article by Goss (1980). The errors associated with moisture determination in com have been the subject of research (Fox and Federson, 1978), and the saponification method of Hood et al. (1971) seems to be a valid method to use with fermented feeds.

Common methods of dry matter determination and suggested occassions for their use are listed in Table1. Students also should check the AOAC (1995) publication for additional methods.. The routine procedure for 100°C oven dry matter determinations is attached to this chapter.

Common methods of dry matter analysis and suggested occasions for their use

Method	Occasions for use
100°C drying	Most mixed feeds, hays, range grasses with 85 to 99% DM
Freeze drying[a]	High-moisture, fermented feeds
Saponification	High-moisture, fermented feeds
Vacuum drying[b]	Meat or tissue samples

[a]Byers (1980) indicated that freeze drying does not yield significantly different values than saponification. See 1. Anim. Sci. 51: 15 8.
[b]Recommended by AOAC for meat samples

Ash is most commonly performed by burning the sample at 550 to 600°C in a muffle furnace. As with the dry matter analysis, some problems can be encountered in that certain minerals (e.g., Se, Pb, Cd) may be volatilized by high-temperature ashing. In cases when this might be important (usually when analysis for these minerals is going to be performed) a procedure known as wet ashing is used. Wet ashing involves digestion of the sample organic matter with nitric and perchloric acids. This procedure, however, is quite dangerous because of the use of perchloric acid, and special precautions such as a perchloric acid fume hood are required. Note that total ash content is not a particularly meaningful analysis, in that it reveals nothing about the content of specific minerals. Moreover, ash will contain any contaminating materials like sand or soil that may inflate a sample's ash value. Generally, ash is a preparatory step for further analysis of specific minerals by spectrophotometric or atomic absorption techniques. The general procedure for ash analysis as well as a procedure for putting ash samples in solution for further analyses is attached to this chapter.

Materials

Aluminum weighing pans, Forced-air oven, Muffle furnace

Procedure

1. Dry aluminum pans in oven at 100°C for 15 to 30 min
2. Cool pans in desiccator, weigh and record weight
3. Add 1- to 2-g samples and record weight of pan plus sample
4. Dry pan plus sample in oven at 100°C for 12 h or overnight

5. Cool in desiccator, weigh back, and record weight
6. Place pans plus samples in muffle furnace and ash at 500°C for 3 h
7. Cool in muffle for at least 8 h, then in desiccator, weigh back, and record weight.

Calculations

Pan plus sample weight (before drying) - Pan weight = Sample wet weight

Pan plus sample weight (after drying) - Pan weight = Sample dry weight

% DM = (Dry weight/Wet weight) × 100

Pan plus sample weight (after ashing) - Pan weight = Sample ash weight

%Ash (DM) = (Ash weight/Dry weight) × 100

5.1.2. Preparation of ash for mineral analyses

Procedures

1. Ash 1 to 2 g sample in muffle furnace.
2. Transfer ash residue to 250-mL beaker.
3. Add 50 mL HC1 (1 part HCl + 3 parts water) and add several drops ofHNO_3; bring to a boil under a hood.
4. Cool and filter into a 50- or 100-mL volumetric flask that has been rinsed with dilute acid.
5. Dilute to volume with deionized H_20.

Note: Some minerals may not be soluble in 25% HCI. Check AOAC or other reference manuals for information on specific minerals.

5.1.3. Crude protein determination by kjeldahl method

Along with dry matter and ash analyses, the Kjeldahl procedure is one of the most common analyses performed in the nutrition laboratory. Although not a measurement of protein *per se,* it does provide an estimate of the content of this important nutrient in feedstuffs. Protein always seems to be in short supply, both in human and animal feeds, and an accurate estimate of protein content is vital in formulating diets for optimum animal performance.

The Kjeldahl method dates back to the 1880's when Johan Kjeldahl used sulfuric and phosphoric acids to decompose organic materials. The method can be conveniently divided into three phases: digestion, distillation, and titration.

The digestion phase is the first phase of the analysis and is designed to oxidize organic matter to CO_2 and H_20, while reducing nitrogen to ammonia. Sulfuric acid is employed as the principle method of decomposing organic matter. Phosphoric acid also can be used in conjunction with sulfuric acid, but an exact mixture of the two acids is critical, and usually the disadvantages of using two acids outweigh the advantages. A number of catalysts have been employed in the digestion phase. Potassium sulfate is almost always added to raise the boiling point of H_2S0_4 and thereby increase rate of digestion. A number of other catalysts have been used to assist in reduction of nitrogen including mercuric oxide, copper sulfate, and selenium. Mercuric oxide and copper sulfate seem to be the most popular and along with potassium sulfate, can be purchased in individually packaged catalyst packs. Because of environmental concerns, the use of mercury-based catalysts has decreased in recent years. At the termination of the digestion phase, organic matter has been decomposed to CO_2 and H_20, and nitrogen has been changed to ammonium sulfate. Digestion requires a temperature of about 350°C and about 2 to 3 h for most feed samples. A good thumb rule is to boil the vessel until contents become clear and then boil for an additional 30 min to 1 h. We should note at this point that nitrates are not completely recovered in most standard kjeldahl procedures. Should one wish to recover all the nitrate-nitrogen in a sample, the addition of salicylic acid as a reducing agent in the reaction mixture will accomplish the task.

The titration process is simply a matter of neutralizing the collected NH_4 +OK with a standard acid. Normally, either HCl or H_2S0_4 is used for this portion of the procedure, and the acid is formulated so that 1 mL of acid will neutralize 1 or 2 mg of nitrogen. Derivation of 0.1428 N HCl, which will titrate 2 mg of nitrogen/mL of acid is attached to this chapter. Rather than preparing HCl of a known normality, it is often more convenient to purchase standardized acid from commerical sources (e.g., 1 mL of .IN HCl will neutralize 1.4 mg of ammonia-N). Once the milliliters of acid used in the titration process are known, the nitrogen and crude protein content can be calculated. Generally, samples to be analyzed for kjeldahl nitrogen are weighed and wrapped in filter paper and then placed in the reaction vessel. Containing the sample in filter paper is simply a method of preventing excessive foaming encountered when loose, ground samples are subjected to H_2SO_4 treatment in the digestion phase. However, because the filter paper contains some nitrogen, it is essential to run a filter paper blank (filter paper + H_2SO_4 + catalysts) to correct for any added nitrogen.

The analysis of ammonia to represent N content and then to express as protein is called "crude" because it does not necessarily represent a amino nitrogen or true protein but may contain all types of non-protein nitrogen (NPN) like urea, amides, nucleic acids, and free amino acids. Generally, for the ruminant this inclusion of NPN is not a major concern because these compounds as well as true protein would be partially converted to ammonia by ruminal microorganisms and subsequently used for microbial protein synthesis. If knowledge of the true a amino-linked (protein) content of a feed is desired, other methods should be used. Students should consult the AOAC manual for further information on this subject. In general, the kjeldahl analysis provides a fairly good estimate of the true protein content of most mixed feeds, hays, grains, and seed meals. Problems are encountered with lush pasture crops and fermented feeds in that a fairly large proportion of the total nitrogen may be present as NPN. The final question is "where does the 6.25 factor in the calculations come from?" Most original work with proteins showed that they averaged around 16% nitrogen. Thus, if nitrogen content is known, one can simply multiply it by 100/16 = 6.25 to estimate protein content. Unfortunately, all proteins are not 16% nitrogen, and in cases where divergence from the 16% value is known, other factors should be used. For example, the combined proteins of milk contain approximately 15.7% nitrogen, and a factor of 6.38 should be used (Maynard et al., 1979). Students should consult Maynard et al. (1979) for further reading and references on this topic.

Kjeldahl N - Macro method

Sample Size

A. Feeds, feces, and dry specimens - weigh a one (1)-g sample and transfer into #1 ashless filter paper, folded to prevent loss of sample.

B. Urine - measure 5 mL of urine sample into an 800 mL kjeldahl flask. Be sure that the specimen is at room temperature and uniformly suspended.

Catalyst: It contains 10 g ofK_2SO_4 plus 0.30 g of$CuSO_4$. For certain procedures it may be preferable to use 15 g of anhydrous K_2SO_4 plus 0.7 g of HgO

a. *Blanks* - Kjeldahl reagents generally contain small amounts of nitrogen, which must be measured and corrected for in calculations. Prepare "blanks" for dry samples by folding one sheet of #1 ashless filter paper and placing it into the kjeldahl flask. Treat blanks exactly like samples to be analyzed, and subtract the amount of acid titrated for the blanks from the amount of acid titrated for samples.

b. Standards - weigh out two (2) .1-g samples of urea, transfer into #1 ashless filter paper and treat exactly like samples. Calculate percent recovery of nitrogen from urea standards.

C. *Acid* - add reagent grade, concentrated H_2SO_4 to kjeldahl flask. For dry samples (feeds and feces) of I-g size, add 25 mL; for dry samples of a 2-g size, add 35 mL; for urine and aqueous samples, add 25 mL. The ratio of acid to dry matter is important, as the boiling temperature is affected by the ratio of acid to salts in the mixture. Also, if too much acid is used, violent reactions can occur at a later step when alkali is added.

D. *Digestion* - tum blower on for digestion rack and burners on '5' setting. After adding sample, catalyst, and acid to flask, place on burners of digestion rack. Rotate flasks about every 15 min to speed the oxidation of carbon. If excessive foaming occurs, cool flask and add about .5 g of purified paraffin wax. After all samples are clear, increase heat to 'Hi' setting and digest for 30 more min. Tum heat off and allow flasks to cool. Do not remove flasks from burners and do not tum blower off until flasks are cool to the touch.

E. *Dilution* - when flasks are cool to touch, add 400 mL of deionized water. Swirl contents to dissolve salts - this is very important!

F. *Preparation of receiving flask for steam distillation* - add 75 mL of prepared boric acid solution (2.9% boric acid containing methyl purple indicator) to a clean 500-mL Erlenmeyer flask and place on distillation rack shelf. Place delivery tube from condenser into the flask, making sure that the tube extends well below the surface of the boric acid solution. TURN THE WATER TO THE DISTILLATION SYSTEM ON, and tum on all burners on the '4' setting

G. Preparation of sample for distillation - Step 1 must be done under the hood as toxic H_2S is given off upon the addition of zinc.

a. Add approximately .5 g of powdered zinc to flask, mix thoroughly and allow settling.

b. Add a scoop of boiling stones.

c. After digest has settled, measure 100 mL of saturated, aqueous NaOH (50% wt/vol) into a graduated cylinder. Slant kjeldahl flask containing prepared digest solution about 45° from vertical position. Pour NaOH slowly. into flask so that a layer forms at the bottom. Do not mix.

d. Attach flask to distillation-condenser assembly. DO NOT MIX flask contents until firmy attached. Holding flask firmly, making sure cork is snugly in place, swirl contents to mix completely. Immediately set flask on heater. Withdraw receiving flask from distillation-condenser delivery tube momentarily to allow pressure to equalize and prevent back suction.

H. Distillation

a. Continue distillation until approximately 250 mL of distillate has been collected in receiving flask.

b. Tum heater off. Remove receiving flask partially and rinse delivery tube with deionized water, collecting the rinse water into receiving flask.

Replace receiving flask with a beaker containing 400 mLs of deionized water. This water will be sucked back into the kjeldahl flask as it cools, washing out the condenser tube.

I. Titration - titrate green distillate back to original purple using. 1 N HCI, and record volume of acid used in titration.

J. Cleaning of flasks - after water has been sucked back into kjeldahl flasks, allow to cool and then pour liquid into sink, catching the boiling stones in a beaker. Wash flasks, INSIDE AND OUTSIDE!!, with tap water, then rinse with dilute acid, followed by deionized water.

Place upside down on kjeldahl flask racks to dry. Make sure you wash both the inside and outside of flasks.

K. Cleaning of area - entire work area should be sponged clean and all reagents, supplies, and glassware returned to proper storage. MAKE SURE that all countertops are free of any spilled acid, alkali, or other reagents!!

L. NOTE - ADIN samples are to be digested on '3' setting for the entire digestion - do not turn to 'Hi' setting. Dilute with 500 mL of deionized water.

Reagents

A. Mixed Indicator - dissolve .3125 g methyl red and .2062 g methylene blue in 250 mL of 95% ethanol. Stir for 24 h.

B. Boric Acid Solution - dissolve 522 g U.S.P. boric acid in 18 L of deionized water. Add 50 mL of mixed indicator solution and allow to stir overnight

C. Zinc – (metallic) powdered or granular, 10 mesh.

D. Sodium Hydroxide - 50% wt/vol aqueous (saturated) technical grade is adequate.

E. Standardized N Hydrochloric Acid Solution - purchased as such from any major chemical supplier.

Digestion

1. Place sample and Kel-pac catalyst into labeled kjeldahl flask.
2. 2.Add 25 mL of concentrated H_2S0_4 for one gram samples or aqueous samples. Add 35 mL of concentrated H_2S0_4 for two gram samples.
3. Tum blower on, and set burners on '5'.
4. Digest until clear, then set on 'Hi' for 30 more min. Rotate flasks occasionally throughout digestion.
5. Cool digest and add 400 mL deionized water.
6. Swirl flask to dissolve salts

Distillation

1. Add 75 mL of boric acid solution to a 500 mL Erlenmeyer flask and label appropriately.
2. Place the delivery tube from the condenser into the boric acid.

3. Tum heater off. Remove receiving flask partially and rinse delivery tube with deionized water, collecting the rinse water into receiving flask

4. Replace receiving flask with a beaker containing 400 mLs of deionized water. This water will be sucked back into the kjeldahl flask as it cools, washing out the condenser tube.

Operation of Tecator rapid digestor and steam distillation unit for Kjeldahl N analysis

Digestion

1. Turn main power switch located on front of digestion controller to ON position. Set temperature to approximately 400°C using temperature set on controller. Allow time for unit to reach temperature.

2. Carefully insert the digestion tubes into the holes provided in the flask rack. A digestion tube must be set in all positions (even if empty) for the exhaust manifold to work properly.

3. Weigh approximately .5 to 1 g of ground homogenous sample in a piece of weighing paper.

   Fold the paper carefully so that the sample is well contained. Place the folded weigh paper into digestion tube, and add 2 Kjeltabs to each tube. Also, weigh a .1 g of urea standard and treat as above. The blank consists of a piece of folded weigh paper and Kjeltabs only. Run one urea standard and one blank with each rack of samples you analyze.

4. When the digestion unit has been on for an adequate warm-up period, add 15 to 20 mL of concentrated sulfuric acid to each tube.

5. After completing sample and reagent addition to the digestion tubes, carefully elevate the rack by grasping the handles provided and place in position on the digestor unit. As the rack is lifted, each tube will rest on its top rim. Carefully lower rack so that each tube enters its respective hole and bottoms in the base of the unit.

6. Place end plates on the flask holder rack, and place the exhaust manifold on top of the tubes.

Turn the water supply on for the exhaust manifold to the highest rate of water flow. *Ensure that the water aspiration-fume removal system is functioning properly before use.*

7. Digest samples at 400°C for approximately 1 h, or until the samples have a clear, blue-green appearance. *Turn down the water supply to exhaust manifold to a lower flow rate after the first* 5 *to* 10 *min of digestion.*

8. After digestion is complete, remove flask rack containing tubes (with fume removal system still attached) from digestor and place in the rack and tray system next to the digestor. Avoid having the hot tubes come into contact with a cool or wet surface. As a safety precaution, wear gloves and goggles when removing the rack and tubes from digestor.

9. Allow tubes to cool for 15 to 20 min. When tubes have cooled, remove the flask rack containing tubes from the fume hood. Place the fume removal system on the rack provided under the fume hood, and tum off the water supply to the fume removal system. Dilute the digest with 100 mL of deionized water when the tubes are cool enough to handle. Dilution must be made before a cake or gel is formed, but not before the digest is cool enough to contain the exothermic reaction. Make sure digest is fully dissolved.

Distillation

1. Turn on the cooling water to the Steam Distillation Unit and allow it to flow through the entire unit. The steam generator drain line should be closed for distillation of samples, but water should flow freely from the waste drain line. The system will automatically adjust water flow to the steam generator.

2. Place the distillation vessel with digested sample on the apparatus, twisting the tube to ensure a complete seal.

3. Insert receiver 250-mL Erylenmeyer flask containing 25 mL of boric acid (Kjel-Sorb) indicator, making sure that the delivery tube is fully submerged.

4. After closing the plastic safety shield door, add approximately 50 mL of NaOH by slowly pushing down the alkali dispenser handle, and allow the dispenser handle to return to its closed position (the dispenser handle is spring-loaded and will return without applicationof force to the handle).

5. Open the steam valve to start steam generation. If desired, set the time for approximately 6 mm.
6. Collect approximatley 175 mL of distillate in the receiver flask. For the last 15 to 25 mL of collection, carefully lower the receiving flask so that the tube is above the fluid level in the receiving flask.
7. Close the steam valve to halt steam generation. Raise the plastic safety shield door, and remove the receiver flask. Use a small quantity of distilled water to was down the outside of the receiver tube into the receiver flask.
8. Wearing gloves, remove the distillation vessel and place it in the nearby tube rack. After vessel has cooled for 10 or 15 min, carefully pour the contents down the drain, diluting with a large quantity of running tap water.
9. To distill another sample, repeat Steps 2 through 8.
10. Titrate the receiving solution.
11. When you have finished distilling and titrating all samples, place a digestion tube that contains approximately 75 to 100 mL of deionized water in the apparatus, start steam generation, and collect 150 to 175 mL of distillate - DO NOT ADD ALKALI FOR THIS STEP. This serves to wash out the system. If needed, repeat this step until the contents of the digestion flask are clear. Close the steam generation valve, tum off the water supply, and tum off the power on the distillation unit.
12. Use a wet paper towel or Kimwipe to wipe the rubber adapter to which the digestion tubes fit, and wash the removable tray in the bottom of the distillation unit. Check inside the distillation unit for other areas that need to be wiped clean. After power has been off to the distillation unit for 15 min, open the drain line for the steam generator.

5.1.4. Extraction and estimation of oil

Estimation of oil in oilseed is generally carried out using the following two methods:

A. Soxhlet method
B. Cold percolation method

Soxhlet method (Method A)

Principle

The principle is based on the extraction of oil using non-polar solvents viz, ether, hexane, petroleum ether (40-60°C). It involves repeated extraction of the material with the solvent. The solvent is then distilled off completely. The oil is dried to a constant weight and % of oil is calculated.

Apparatus

1. Soxhlet extractor assembly
2. Cotton wool
3. Whatman filter paper/Thimble.

Reagents

Hexane/diethyl ether/petroleum ether (40-60°C)

Procedure

1. Fold a piece of filter paper and make into a sample packet in such a way to hold the finely crushed weighed seed material (2-5g depending on oil content). Alternatively, the sample can be transferred directly into a thimble.
2. Place the sample packet into extractor of Soxhlet apparatus, after placing some cotton at the bottom. A piece of cotton is placed at the top to evenly distribute the solvent as it drops on the sample during extraction.
3. Add organic solvent, two and half times the capacity of the extractor in the flask and extract oil for a period of six hours or for a longer period till the solvent in the extractor becomes colourless.
4. Put off the heaters and allow to cool. Flash evaporate the solvent or distill is completely. Finally, it is heated on water bath to remove last traces of the solvent. Keep it in oven at 70°C for 10 min. Cool at room temperature.
5. Weigh the flask after removal of moisture, if any Repeat heating until constant weight is recorded.

Calculation

Weight of the oilseed	=	Wg
Weight of flask + glass beads	=	Xg
Weight of flask + glass beads + oil	=	Yg
Weight of oil	=	(Y - X)g

$$\% \text{ Oil} = \frac{(Y-X)}{W} \times 100$$

Precautions

1. Put off flame in the lab while extracting oil using organic solvents.
2. To check bumping use glass beads in the flask.
3. The oilseed material should be finely crushed to extract oil from it.
4. To remove moisture, if any, the extractant should be treated with some anhydrous sodium sulphate.

Cold percolation method (Method B)

The extraction of oil from the sample involves the use of non-polar solvents viz. hexane, petroleum ether (40-60°C) which percolate through a column containing finely powdered oilseed along with anhydrous sodium sulphate. The oil gets eluted by the solvent and is collected in a conical flask.

Reagents

(i) Anhydrous sodium sulphate

(ii) CCl_4 / hexane / petroleum ether (40-60°C) / solvent ether.

Procedure

(i) Exactly weigh (1-2 g) ground seed material along with 20-30 g of anhydrous sodium sulpahte, mix well.

(ii) Plug the percolator by putting cotton/glass wool at the bottom of it. Put a layer of approximately 2 g of sodium sulpahte (anhydrous) over the plug.

(iii) Pack the powdered material slowly inside the percolator carefully. Ensure that whole of The material is transferred properly. Pack little cotton over the material and cover with another layer of anhydrous sodium sulphate.

(iv) Add 10 ml of solvent in the mortar and pestle to ensure complete removal of the material and transfer it into the percolator.

(v) Keep a flask below the percolator and fill the percolator with the solvent.

(vi) Let the solvent tickle down under gravity. The solvent will flow down extracting oil from the material.

(vii) Add more solvent to fill the percolator and allow it to pass through the column. Repeat this thrice.

(viii) Flash evaporate or the solvent evaporate on water bath. When 5 ml of solvent is left in the flask (Round bottom in case of flash evaporator/distillation), remove it and transfer it into a weighed conical flask and keep it in the oven till the solvent is evaporated. Weigh to a constant weight after keeping in the desiccator for few minutes.

(ix) Calculate oil per cent as given above (Method A).

Precautions

1. Put off flame in the lab while extracting oil using organic solvents.
2. The oilseed should be crushed finely along with anhydrous sodium sulphate.
3. Column should not be packed too tight or too loose.
4. Use glass beads in the flask while evaporating the solvent to check bumping.
5. Add some anhydrous sodium sulphate to the extract to remove last traces of moisture and then filter it off.

Reference

Kartha ARS and Sethi AS (1957). Ind. J. Agric. Sci. **27** : 211. Association of Official Analytical Chemists (1965). Methods of Analysis.

5.1.5. Estimation of fibre content of feedstuffs

Crude Fibre

Crude fibre consists of cellulose, variable proportion of hemicellulose and highly variable proportion of lignin alongwith some minerals. The estimation is based on treating the moisture and fat free sample successively with dilute (1.25%) acid and alkali.

Reagents

(i) 1.25% H_2SO_4 (W/V) solution.

(ii) 1.25% NaOH (W/V) solution.

Procedure

1. Weigh about 2g moisture and fat free sample and transfer it to the one litre spoutless beaker.
2. Add 200 ml of 1.25% sulphuric acid.
3. Place on hot plate, cover the beaker with round bottom flask condenser and allow to reflux for 30 minutes after it has started boiling.
4. Shake the content every five minutes.
5. After boiling for 30 minutes, remove the beaker from hot plate and filter through muslin cloth (should have 18 threads per cm) using suction. Work the residue with hot water till it is free from acid.
6. Transfer the material to the same beaker and add 200 ml of 1.25% NaOH solution.
7. Again reflux the contents for 30 minutes.
8. Filter again through the muslin cloth with the help of vaccum pump and wash the residue with hot water till it is free from alkali (A wash of residue on muslin cloth with dilute HCl facilities removal of alkali before washing with hot water facilitates removal of alkali).
9. Transfer the total residue to a crucible and place it in hot air oven, allow to dry to a constant weight at 80^0 -110°C and record its weight.
10. Ignite the residue in muffle furnace at 550°-600°C for 2 to3 hrs., cool and weigh again. The loss of weight due to ignition is the weight of crude fibre.

Calculations

$$\text{Crude fibre}(\%) = \frac{\text{Weight of crude fibre}}{\text{Original weight of sample}}$$

The method described above for crude fibre determination does not give the exact value of the fibre because some of the constituent like cellulose, hemicellulose and lignin are partially retained in the crude fibre fraction.

5.2. Detergent Fibre Analysis

P. J. Van Soest and associates, working at the USDA station at Beltsville, MD, developed a rapid technique of separating feed carbohydrates on the basis of nutritional availability to ruminants and ruminal bacteria. Essentially, the method divides feeds into two fractions: (1) plant cell contents, a highly digestible fraction consisting of sugars, starches, soluble protein, pectin, and lipids; (2) plant cell wall constituents, a fraction of variable digestibility consisting of insoluble protein, hemicellulose, cellulose, lignin, and bound nitrogen. The method involves boiling a sample in a neutral detergent solution. The soluble fraction is termed neutral detergent solubles (cell contents), whereas the fibrous residue is called neutral detergent fiber (cell wall constituents). Unlike crude fiber and NFE, both NDS and NDF accurately predict the proportions of more and less digestible fractions, respectively, found in a wide variety of feedstuffs.

The Van Soest (or detergent) scheme has been further refined with the addition of acid detergent fiber (ADF) analysis, which breaks down NDF into a soluble fraction containing primarily hemicellulose and some insoluble protein and an insoluble fraction containing cellulose, lignin, and bound nitrogen. Furthermore, the content of lignin in ADF can be determined by either treating the fiber with H_2S0_4 to dissolve the cellulose or by oxidation with permanganate to degrade the lignin. This analysis is quite important because lignin has been shown to be a major factor influencing the digestibility of forages. As indicated in the discussion of the Weende system, two important reasons for development of the system are so that comparisons of nutrient content and prediction of animal performance could be made. The same reasons apply to the detergent scheme; fortunately, the detergent scheme does a better job than the Weende system in both respects.

5.2.1. Acid Detergent Fiber (ADF) Procedure

Materials

600 mL Berzelius beakers Fiber digestion apparatus, Sintered glass crucibles - 40 to 50 mL - coarse porosity Filtering flasks

Reagents

Acid Detergent Solution: Add 27.84 mL ofH_2S0_4 to a volumetric flask and bring to 1 L volume with H_20. Then and 20 g of cetrimide $[CH_3(CH_2)lsN(CH_3)_3Br]$ to the 1 L of acid solution.

Digestion of Sample

A. Transfer 1 g air-dried sample to Berzelius beaker

B. Add 100 mL acid detergent solution.

C. Heat to boil (5 to 10 min), and boil exactly 60 min.

D. Filter with light suction into previously tared crucibles. Refer to procedure for matting crucibles. #541 filter paper may be used instead of crucible if ADL is not being done. Must preweigh filter paper first.

E. Wash with hot water 2 to 3 times.

F. Wash thoroughly with acetone until no further color is removed. Suction dry

G. Dry in drying oven at 100GC for 24 h

H. Cool in dessicator. Weigh and record weight.

5.2.2. Indigestible ADF Procedure

1. Run 96-h in vitro and 48-h acid-pepsin digest on feed, digesta, or feces (use standard in vitro procedure as outlined in this laboratory manual with .5-g samples).

2. Empty tube contents into Berzelius beaker (without centrifugation), and rinse tube with ADF solution. Add a total volume of 100 mL of ADF solution.

3. Run standard ADF procedure (Whatman #541 filter paper may be used instead of tared crucibles) .

% Indigestible ADF = (({ dry residue - [pan + filter paper wt]) - {blank - [pan + filter paper wt]})/ dry sample wt) x 100

5.2.3. ADL Procedure

Reagents

A. 72% H_2S0_4 standardized to specific gravity of 1.634 at 20^{G}C.

Place ADF crucible in a 50 mL beaker on a tray.

B. Cover contents of crucible with 72% H_2SO_4. (Fill approximately half way with acid.)

Stir contents with a glass rod to a smooth paste.

D. Leave rod in crucible, refill hourly for 3 h, each time stirring the contents of the crucible.

E. After 3 h, filter contents of crucible using low vacuum at first, increasing it only as more force is needed.

Wash contents with hot water until free of acid (minimum of five times).

Rinse rod and remove.

Dry crucible in oven at *100°C* for 24 h.

Cool in dessicator. Weigh and record weight.

Ash in muffle at *500°C* for 4 h.

Cool in dessicator. Weigh and record.

5.2.4. ADIN Procedure

A. Run ADF on sample just as described in procedure. Filter through #541 filter paper.

B. Place contents of filter paper and residue from ADF procedure in kjeldahl flask. A blank for this procedure would consist of a #541 filter paper taken through ADF procedure.

C. Follow procedure for determining Kjeldahl nitrogen, and do calculations on original sample weight.

5.2.5. Neutral Detergent Fiber (NDF) Procedure

Reagents and Materials

NDF solution:

To 1 L of H_2O add:

30 g of sodium lauryl sulfate

18.61 g of disodium dihydrogen ethylene diamine tetraacetic dihydrate 6.81 g of sodium borate decahydrate

4.56 g of disodium hydrogen phosphate

10 mL of triethylene glycol

Amylase solution - Heat-stable a-amylase (Sigma No. A3306 - from the Dietary Fiber Kit). Acetone

Refluxing apparatus Whatman #541 filter paper Aluminum pans

Procedure

1. Place 0.5 to 1.0 g sample in 600-mL Berzelius beaker.
2. Add 100 mL of neutral detergent fiber solution.
3. Heat to boiling (5 to 10 min). Decrease heat as boiling begins. Boil for 60 min.
4. After 60 min, filter contents onto preweighed Whatman #541 filter paper under vacuum. Use low vacuum at first, increasing only as more force is needed.
5. Rinse contents with hot water, filter, and repeat twice.
6. Wash twice with acetone.
7. Fold and place in preweighed aluminum pan.
8. Dry overnight in 100ac oven.
9. Cool in dessicator. Weigh - yield is CWC.
10. For samples with a high starch content: Add 50 ul. of heat-stable amylase to the beaker along with NDF solution as in Step 2, and follow remaining steps. For the most difficult samples, a I-g sample can first be treated with 30 mL of 8 M urea solution plus 50 ul. of heat-stable amylase solution. The mixture can be heated on a steam bath at 80 to 90°C for 5 min, then incubated at room temperature for 4 h or overnight. After incubation, add 100 mL of NDF solution and treat as in Step 3 and following. An additional 50 ul. of heat-stable amylase can be added at this point if desired.

Reference

Goering and Van Soest (1970), as modified by Van Soest et al. (1991; J. Dairy Sci. 74:3583).

5.2.5.1. Indigestible NDF

1. Run 96-h in vitro and 48-h acid-pepsin digest on feed, digesta, or feces (use standard in vitro procedure as outlined in this laboratory manual with .5 g samples).

2. Empty tube contents into Berzelius beaker (without centrifugation), and rinse tube with NDF solution. Add a total volume of 50 mL ofNDF solution.
3. Run procedure for slow-filtering NDF, (i.e., use amylase). Whatman #541 filter paper is used for filtration.

% Indigestible NDF = (({ dry residue - [pan + filter paper wt]) - {blank - [pan + filter paper wt]})/dry sample wt) x100

5.2.6. Procedure for Matting Crucibles

1. Sonicate crucibles for 2 min.
2. Reverse flush with 50 mL of dilute acid.
3. Rinse with hot water.
4. Add a teaspoon of Hiflo supercell.
5. Add 30 mL of hot water, swirl and place on sidearm flask.
6. Pour 100 mL of dilute acid through crucible.
7. Pour 100 mL of hot water through crucible.
8. Place in muffle *(500°C)* for 4 h.
9. Cool in dessicator, record tare weight.
10. After analyses, discard mat.
11. Wash, then sonicate for 2 min.
12. Backflush with 50 mL of dilute acid and 50 mL of hot water.

5.3. Energy in Animal Nutrition

Oxygen Bomb Calorimetry

The bomb calorimeter consists of the bomb, an airtight canister in which the substance is burned, and a jacket of H_20 around the bomb that provides a means of measuring the heat produced during burning. In an adiabatic calorimeter, an additional water jacket surrounds the inner jacket. This second water jacket is maintained at the same temperature as the inner jacket to prevent heat loss by conduction. Thus, when a substance is burned in the bomb, the temperature rise in the inner jacket surrounding the bomb reflects the heat produced.

When a substance is completely burned to its ultimate oxidation products of CO_2 and H_2O or other gases, the heat given off is termed gross energy or heat of combustion. Gross energy is generally measured in an oxygen bomb calorimeter and is the starting point for determining the energy value of feedstuffs (Maynard et al., 1979). The method involves placing a small sample (usually not more than 1 g) in a cup inside the bomb. A fuse wire is attached to terminals in the bomb and placed over the sample. The bomb is then sealed and charged with 20 to 25 atmospheres of oxygen. Then the bomb is placed in the inner water jacket, which contains a known weight of water. When the inner water jacket and outer water jacket have reached the same temperature, a charge is sent to the fuse wire, and the sample bums rapidly in the oxygen-rich environment. Heat from the burning is transferred (in part) to the inner water jacket and the increase in temperature is recorded. Once the change in water temperature (~T) is known, the gross energy of the substance can be calculated by multiplying the thermal equivalent of the calorimeter by ~ T. Every calorimeter has its own thermal equivalent, which is determined by running a standard of known energy content and determining the calories per °C rise in temperature. Once a calorimeter's thermal equivalent has been determined, it should remain the same, provided none of the parts are changed. However, in practice, the thermal equivalent is determined periodically as a check on the instrument and the operator's technique. Normally, benzoic acid (supplied in preformed pellets by Parr Instruments) is used in determining the thermal equivalent. Some corrections to the gross energy value are necessary because the fuse wire produces some heat when it is burned and nitrogen and sulfur are oxidized to NO-3 and SO-4 under high O2 tension. Because such oxidation would not occur under physiological conditions, the heat generated in the process is subtracted from the determined gross energy value. If we were to run a number of purified nutrients through gross energy determinations, we would find that on the average, fats contain about 8.9 to 9.6 kcal/g of GE, whereas carbohydrates contain 3.7 to 4.4, and proteins contain about 5.4 to 5.9 kcal/g.

Why are there such large differences?

These differences are governed by their (nutrients) elemental composition, especially by the relative amount of oxygen contained within the molecule, since heat is produced only by the oxidation resulting from the union of carbon or hydrogen with oxygen from without. The oxygen within has previously liberated heat (or energy) during chemical formation of the compound. **In** the case of carbohydrates, there is enough oxygen in the molecule to take care of all the hydrogen present, and thus heat arises only from the oxidation of the carbon. In the case of fat, however, there is relatively much less oxygen present and relatively more atoms requiring oxygen from without, and the combustion involves the oxidation of hydrogen as well as carbon. The burning of 1 g of hydrogen produces over four times as much heat (34.5 kcal/g) as is the case of carbon (8 kcal/g). These facts explain the much greater gross energy values per gram for fats compared with the carbohydrates. The heat produced in the burning of protein comes from the oxidation of both carbon and hydrogen, but the nitrogen present gives rise to no heat at all because it is set free as such in its gaseous form. No oxidation of it has taken place, and thus no heat is produced, contrary to what occurs in the bomb, for which partial corrections are made. Simply stated, nutrients vary in energy content because they vary in degree of internal oxidation, fats being less internally oxidized than carbohydrates. Students can readily see why the inclusion of a small amount of fat in an animal's diet markedly increases the energy content.

5.4. Nitrogenous Constituents of Feeds

For many years, the Kjeldahl analysis has been the standard method of evaluating the nitrogen content of feedstuffs. However, in recent years, more attention has been given to the availability of feed nitrogen. It has long been recognized that the total nitrogen content of feeds could be divided into non-protein nitrogen and protein nitrogen fractions. Moreover, a portion of the true protein may be bound or unavailable as a result of excessive heating of the feed or other treatments and, in the case of ruminants, nitrogen is either insoluble or soluble in ruminal contents. With all these variations, the need for a clearer description of feed nitrogen is evident.

Soluble protein is consists of true protein and various types of non-protein nitrogen (NPN), including amino acids, nucleic acids, nitrates, amines, urea, and so on. Soluble protein is considered to be rapidly degraded to ammonia by ruminal microorganisms. In fermented feeds, a large portion of the soluble protein may be in the form of non-protein nitrogen and, hence, it is desirable to divide the soluble protein fraction into soluble NPN and soluble true protein. A procedure for separating soluble protein into these two fractions has been described in literature. Essentially, this involves placing the sample in a buffer solution and agitating for 1 h at 39°C. The sample is then filtered and the filtrate divided into two portions. One portion is analyzed by kjeldahl and the total N x 6.25 is an estimate of soluble crude protein. The second portion is treated with tungstic acid to precipitate proteins, filtered and the filtrate analyzed for N by kjeldahl, yielding an estimate of soluble NPN. Total soluble nitrogen minus soluble NPN is an estimate of soluble true protein nitrogen. The procedure for soluble and insoluble N, based on the .15 M NaCl procedure of Waldo and Goering (1979), is attached to this chapter.

Insoluble available protein has been the subject of considerable research in ruminant nutrition. This fraction consists largely of true protein with small amounts of NPN and is essentially insoluble in ruminal fluid and degraded to a variable extent by ruminal microorganisms, depending on the nature of the protein. Because it is only partially degraded in the rumen, insoluble protein is often considered as the portion of total protein with escape potential. Escape of feed protein could be important in ruminants, particularly when the escape protein is of greater quality (biological value) than microbial protein. A number of methods of treating protein to increase ruminal escape have been studied. Formaldehyde treatment, heating, and coating the protein with oil have all met with some success.

Before we move on, it should be noted that knowledge of the solubility and insolubility of protein in buffers or ruminal fluid is of limited value in describing feeds for non-ruminants. Perhaps a more common approach would be to evaluate protein in terms of its digestibility in pepsin. Pepsin indigestible protein would provide a fairly good index of that portion of the total protein which would be unavailable to non-ruminants. It also appears, however, that pepsin indigestible protein is not digested in ruminants either. A procedure for determination of pepsin indigestible protein is in the AOAC (1980) manual.

Finally, let us consider the bound protein fraction. This fraction is essentially the undigestible portion of the feed nitrogen. Components of the bound fraction include naturally occurring nitrogen in lignin, protein-tannin complexes, and proteins or amino acids bound to carbohydrates by the Maillard reaction. Perhaps the most important of these three from the standpoint of feed preparation is the Maillard, or browning reaction. **In** this reaction, free amino groups of the protein (usually the s-amino of lysine) combine with the aldehyde group of a reducing sugar to form an indigestible complex. The reaction is maximized at moisture levels of 30 % and temperatures greater than

60°C. Moreover, increased time of exposure to these conditions will enhance the reaction (Maynard et al., 1979). The complex formed by this reaction is measured as lignin in the ADL analysis and is sometimes referred to as artifact lignin. The Maillard reaction is of considerable interest to the protein processing industry, particularly with regard to the effects of heat on protein quality. Proteins that have undergone browning have considerably decreased value, both for ruminants and nonruminants. Several methods have been developed to measure bound protein. One method involves mixing fluorodinitrobenzene (FDNB) with the protein. FDNB complexes with free amino groups and upon hydrolysis of the protein, the FDNB-amino acid complex produces a yellow color. Thus, the more intense the yellow color, the less the extent of amino acid-sugar.

Pepsin indigestible protein also would provide an index of bound protein; however, the most common procedure with forages is the measurement of acid detergent insoluble nitrogen (ADIN). The ADIN procedure was included with this Chapter. In forages, the principal carbohydrate involved in the Maillard reaction is hemicellulose and the reaction may be a considerable problem in ensiled or artificially dried hays where moisture and temperature conditions are condusive to development of the reaction.

5.4.1. Soluble nitrogen content of feedstuffs

1. Determine total N content of feed sample by Kjeldahl.
2. *Weigh out samples of.5 g into a 125-mL Erlenmeyer flask.
3. Add 50 mL of .15 M NaCl (preheat NaCl to 40°C). The pH of the NaCl should be adjusted to 6.5 with orthophosphoric acid.
4. Place flask in H_20 bath (40°C) for 6 h. Stir occasionally.
5. Filter contents of flask through Whatman #4 filter paper held in a Buchner funnel.
6. *Determine by Kjeldahl amount of N in residue and calculate soluble N as difference between total N and insoluble N.

   *Sample size and volume of NaCl could be increased proportionally for samples low in crude protein content.

   **By determining ADIN on the same sample, the insoluble fraction can be subdivided into insoluble available and unavailable fractions.

References

Carpenter, K. 1. and V. H. Booth. 1973. Damage to lysine in food processing: Its measurement and significance. Nutr. Abst. Rev. 43:424.

Ferguson, K. A. 1975. The protection of dietary proteins and amino acids against microbial fermentation in the rumen. In: McDonald, 1. W., and A. C. 1. Warner (Ed.), Digestion and Metabolism in the Ruminant. Univ. of New England Publ. Unit, Armidale, Australia.

Maynard, L. A., J. K. Loosli, H. F. Hintz, and R. G. Warner. 1979. Animal Nutrition. McGraw Hill Book Company, New York, N.Y.

Van Soest, P. 1. 1975. Physio-chemical aspects of fibre digestion. In: McDonald, 1. W. and A. C. 1. Warner (Ed.), Digestion and Metabolism in the Ruminant. Univ. of New England Publ. Unit, Armidale, Australia.

Waldo, D. R. and H. K. Goering. 1979. Insolubility of proteins in ruminant feeds by four methods. J. Anim. Sci. 49:1560.

5.5. Mineral and Inorganic Constituents of Feedstuffs

5.5.1. Determination of phosphorus

Method: Colorimetric Method - AOAC 2.019, 2.095-7.098

Reagents

Molybdovanadate reagent - Dissolve 40 g Nl-la-rnolybdate AH_20 in 400 mL hot H_20 and cool. Dissolve 2 g Nl-la-metavanadate in 250 mL hot H_20, cool, and add 250 mL 70 % $HC10_4$. Gradually add molybdate soln. to vanadate soln with stirring and dilute to 2 L.

Phosphorus standard solns. - (1) stock soln. - 2 mg P/ mL. Dissolve 8.788 g KH_2P0_4 in H_20 and dilute to 1 L. (2) working soln. - 0.1 mg P/ mL. Dilute 50 mL stock soln. to lL.

Preparation of Standard Curve

Transfer aliquots of working std. soln. containing .2, .5, 0.8, 1.0, and 1.5 mg P to 100 mL volumetric flasks (this corresponds to 2,5,8, 10, and 15 mL of working soln). Treat as below.

Make sure all glassware has been rinsed with dilute acid before use. Place aliquot of sample solution containing .2 to 1.5 mg P in 100-mL volumetric flask. Add 20 mL molybdovanadate reagent, dilute to volume with H_20, and mix well. Let stand 10 min, then read at 400 nm using the 0.2 mg standard as the blank. Determine milligrams of P from standard curve.

5.5.2. Calcium determination

Reagents: Lanthium oxide solution

Add 58.65 g of La_203 to I-L volumetric flask, slowly add 250 mL of HCI under hood, dissolve completely, bring to volume with reagent grade H_20, and cap tightly.

Sample preparation: as in above procedure

- Ash -1 to 2-g sample in muffle in duplicate
- place residue in 250-mL beaker
- Add 50 mL of 1:3 HCI (1 part HCI to 3 parts H_20) and several drops of$HN0_3$; bring to boil under hood
- Cool and filter into 50- or IOO-mL volumetric flask that has been rinsed with dilute acid
- Dilute to volume with reagent grade H_2O

Standard preparation

Place 1 mL of stock Ca solution (1,000 ug/ml.) into IOO-mL volumetric with Hamilton syringe, and bring to volume with reagent grade H_20. This is working solution (10 ug/ml.). Place 0, 2.5, 5, 10,20, and 30 mL of working solution into 100mL volumetrics with acid-rinsed glass pipettes, add 20 mL of La_203 solution, bring to volume with reagent grade H_20, and cap (0 serves as blank; others equate to .25, .5, 1,2 and 3 ug/ml. standards).

Unknown preparation

Place a predetermined aliquot of unknown (ashed sample in solution) into 8-mL culture tube with Hamilton syringe, add 1 mL of La_203 solution and enough H_2O to result in a total volume of 5 mL, and cover tubes

Read standards and unknowns on atomic absorption spectrophotometer (refer to instrument manual for wavelength, etc.). Results are expressed in concentration (mg/mL).

* Calculation of aliquot to use:

Estimate Ca content of unknown

Example

If there is 0.7% Ca in sample, and a I-g sample is used and brought to IOO-mL volume, then 1 g has 0.007 g of Ca (0.007 x 1 g), and 0.007 g = 7 mg of Ca.

7 mg/IOO mL = 0.07 mg/mL = 70 ug/ml.

70 ug/ml. x .1 mL = 7 ug (in this aliquot)

7 ug/S mL = 1.4 ug/ml. in sample to be read. This falls in the readable range of the standards, thus, .1 mL (100 ul.) is an appropriate aliquot. Note: the amount of H_20 to bring to 5 mL would be 3.9 mL. Note: if the sample size is larger (i.e. 2 g) or brought to volume in 50-mL volumetric, Ca will be more concentrated, and a smaller aliquot would be appropriate.

Calculations

AA reading yields ug/ml. in tube.

Multiply this value by 5 to determine the quantity of Ca in the aliquot.

Divide volume of ash in solution by aliquot taken from that solution:

Example - 100/0.1 = 1,000

Multiply amount of Ca in aliquot by this factor to determine quantity of Ca in volumetric flask.

Divide by dry sample weight: (Total ug/g of dry sample x $19/10^6$ ug) x 100 = % Ca, dry matter basis.

5.5.3. Atomic absorption and emission spectroscopy

Atomic absorption and flame emission spectrometry are analytical techniques developed in recent years for the determination of metallic element concentrations. Flame emission involves measurement of the intensity of radiation of an excited atom at a given spectral line, whereas atomic absorption involves measurement of light absorption by free atoms. We will briefly discuss each method and consider their application to mineral analysis.

Flame Emission Spectrometry. When a sample is sprayed into a flame, the following events occur (Fritz and Schenk, 1979):

1. The solvent is evaporated, leaving airborne particles of solids that were dissolved in the solvent.
2. Solid compounds are vaporized and partially converted into gaseous atoms.

   Mg Cb (solid) —+ MgCb (gas) —+ Mg (gas) + CI (gas)

3. A small fraction of the gaseous atoms are excited by the thermal energy. When electrons in these atoms return to the ground state, they emit a photon (hv) ofUV or visible light:

   heat
   Mg (gas) .... + Mg* (gas) —+ Mg (gas) + hv (* = excited state)

If a constant set of conditions is maintained, the light emitted by the atoms will be proportional to the concentration of the element in the sample. In most atoms, a number of electron transitions to higher energy states are possible, each giving rise to light of a specific wavelength or spectral line. The most prominent line, however, is that resulting from an electron returning from the lowest excited state to the ground state, and this line is usually chosen for analysis (Fritz and Schenk, 1979).

Now, let us consider the analytical procedure for determining the concentration of elements by flame emission. Inorganic samples are dissolved in acids as with most spectrophotometric procedures. Standard solutions containing elements of interest are prepared and aspirated into a hot flame, producing some free atoms of the element. These atoms become excited and give off light. A monochromator is set to select light of the spectral line of the element in question and the intensity of emitted light, as indicated by the current from a photomultiplier tube, is plotted against standard concentration. The unknown samples are aspirated into the flame, and their concentration is determined by reference to the standard curve. It should be noted at this point that intensity of emitted light is usually not linear with concentration in flame emission procedures, and polynomial or curvilinear regression techniques are often used to find the equation that best fits the standard data. This problem of non-linearity is the result of self-absorption in the flame, (i.e., energy emitted by an atom is absorbed by another atom of the same element, thereby decreasing the intensity of emitted light). Selfabsorption can be decreased by using dilute solutions.

There are several interferences that can influence the results of flame emission procedures. Most of these affect atomic absorption measurements also, but to different extents. Spectral interferences result from the emission of light by atoms of other elements in the sample that have spectral lines close to the element of interest. Although monochromators remove most of the problems with this type of interference, it is not uncommon to correct flame measurements for background or solvent emissions.

Ionization interference results when metal atoms become ionized and, thus, emission of light at the desired wavelength is decreased. This problem is reduced by adding the salt of an easily ionized element to the samples. An excess of electrons from the easily ionized element reverses the equilibrium of the element of interest, so very few of its atoms ionize (Fritz and Schenk, 1979).

Exitation interferences can be a considerable problem in flame emission measurements. These result from changes in the flame or other conditions that alter the number of atoms in the excited state. It is possible to use an internal standard to adjust for such variations. This involves adding (to all samples) a known amount of another element with a spectral line close to the element of interest. If the emission intensity of the internal standard changes, adjustments can be made to intensity values of the element of interest.

The final type of interference is chemical interference. This results from the formation of salts or oxides in the flame that do not form free atoms. A classic example is the effect of phosphate on calcium, in which a stable calcium phosphate complex is formed that does not emit light. This problem can be overcome by adding lanthanum oxide to samples and standards, which binds phosphate, freeing calcium for exitation in the flame. The Animal Nutrition Laboratory procedure for analysis of Ca by atomic absorption is attached to this chapter. Another chemical interference is the formation of stable metal oxides in the flame. However, oxide formation can be decreased by fuel-rich flames.

Inductively Coupled Plasma Emission Spectroscopy. A recent development in emission spectrometry has been the use of electrically generated plasmas, rather than fuel-fired flames to produce atomic emission. A plasma is a gas in which a significant number of the atoms or molecules are ionized. Because it is an electrical conductor, it can be heated rapidly by inductively coupling it to a time-varying magnetic field (Fritz and Schenk, 1979). The temperature reached by the process is two to three times that obtained in routine flame emission. The high temperature seems to overcome many of the interferences in flame emission, and spectral interferences are about the only problem with the procedure. Use of ICP allows for excellent detection limits, and when computer-coupled, 20 or more elements can be analyzed in a matter of minutes. For more information, students are referred to Fassel (1978).

Atomic Absorption Spectrometry. Atomic absorption is essentially the opposite of flame emission. As in flame emission, the sample is converted into free atoms by aspirating it into a flame. In this case, however, a light source emitting the spectral line of the element of interest is used and free atoms in the flame absorb the light. Generally, the great majority of free atoms remain in the ground state in the flame, and are thereby capable of absorbing the light. The amount of absorption is proportional to the concentration of the element in question (i.e., Beer's Law may be applied).

As should be evident, the atomic absorption spectra of free atoms is very narrow, compared with the broad spectra of molecules in solution (Fritz and Schenk, 1979). In comparison with flame emission, atomic absorption has the advantage of fewer interferences. In fact, chemical interferences are the only significant concern in atomic absorption procedures.

The *hollow-cathode lamp* is an evacuated tube containing a noble gas, usually neon or argon. The hollow- or tube-shaped cathode is made of the element to be determined. When voltage is applied, the gas is ionized, and ions bombard the cathode (Fritz and Schenk, 1979), causing release of metal atoms that collide with gas ions. This elevates the metal atoms to an excited state. Upon fall to the ground state, the metal atoms give off light of the desired wavelength.

Generally, the light emitted by a hollow cathode tube is modulated or chopped so that the detector can distinguish it from light emitted by atoms in the flame. If this is not done, absorbance values will seem to be lower than they actually are. Modulation is accomplished by either manually chopping the light beam or alternating the current to the light source and tuning the detector to the same frequency. In either case, the constant emission of light by excited atoms in the flame will not influence the detector.

The flame is equivalent to the defined space or cuvette in general spectrophotometry. Generally there are two types of burners: total consumption or pre-mix. Total-consumption burners do not provide a very steady flame because gas and sample enter at a single opening. This type of burner is not extensively used in newer instruments. Pre-mix burners mix fuel and sample before entering the flame, resulting in a much steadier flame than a total-consumption burner. All newer instruments use pre-mix burners. The nebulizer is simply an aspirating device to draw up the sample and convert it to a fine mist before it enters the flame.

References

Fassel, V. A. 1978. Quantitative elemental analyses by plasma emission spectroscopy. Science 202:183.

Fritz, J. S., and G. H. Schenk. 1979. Quantitative Analytical Chemistry.

5.6. Determination of Silica

Materials

1. 250-mL beakers
2. 6N HCI
3. Volumetric flasks (50-mL or IOO-mL)
4. Funnels
5. Tared crucibles
6. Whatman No. 541 filter paper

Caution : Wear gloves, lab coat, and safety glasses for this procedure (STRONG ACID)

Procedure

1. Weigh out a I-g sample, and ash for 6 h in a 600°C muffle furnace.
2. Place the ashed sample into a clean and clearly labeled IOO-mL Pyrex beaker, and add 50 mL of 6 N HCI (can use 1:3 HCI dilution with water).
3. Place the beakers on a hot place under the hood.
4. Boil off the acid. **CAUTION:** Do not leave your samples unattended, or the beakers can crack.

5. After removing from heat, add 25 mL of 6 N HCI to put the dehydrated ash back into solution
6. Filter the liquid contents of each beaker into a funnel holding a piece of No. 541 filter paper, and collect the liquid into a labeled volumetric flask (save when mineral analysis is to be done). Bring to volume using deionized H_20.
7. Place the filter paper into a previously tared crucible, and ash in a muffle furnace for 6 h at 600°C. (Do not use aluminum pans - they will corrode.)
8. Weigh back residue as SiO_2.

   Calculation : % silica = ({[wt of crucible + ash] - [crucible wt]}/ sample wt) x 100

□□□

Chapter 6

Analysis of Antinutrients and Toxic Principles

6.1. Total Phenolics, Hydrolysable and Condensed Tannins

The methods delineated by Makkar (1994) is followed for analysis of tannin constituents in selected feedstuffs.

6.1.1. Extraction of tannins

Two hundred mg selected feed sample is taken in a 50 ml conical flask. Before extraction of tannins, the pigments and fats from the sample are removed by 20 ml of diethyl ether containing 1% acetic acid. This mixture is shaken thoroughly and kept for 5 min. Diethyl ether is then decanted. After removal of diethyl ether by drying the residue at 30^0 C, 10 ml of 70% aqueous acetone is added to the defatted and depigmented residue. The flask is shaken for 2 h in an orbital shaker (30^0 C, 130 rpm). The contents are centrifuged for 20 min at 5,000 g. The supernatant is collected and stored at 4^{0}C, if not immediately processed for analysis of tannins.

6.1.2. Extraction and purification of condensed tannins

Two grams of leaves powder is taken in each of the five 250 ml flasks and 100 ml 70% aqueous acetone (containing 0.1% ascorbic acid) is added to each flask. The flasks are kept in an orbital shaker at 120 rpm and 30°C for 2 hrs. The suspension is filtered through Whatman filter paper no. 1. The filtered acetone extract is pooled and its volume noted. The acetone extract was taken in a separating funnel and equal volume of dichloromethane is added to remove chlorophyll and lipids. The crude defatted extract is obtained by partitioning the acetone extract with dichloromethane. The crude defatted extract is then lyophilized. For further purification method of Foo *et al.* (1997) was adopted. The freeze dried extract is reconstituted with 50% methanol for column chromatography.

Column chromatography

Twenty-five gram of Sephadex LH-20 is weighed and 50% methanol was added to it. It is then kept overnight to get maximum swollen volume. The gel is poured into the column and is equilibrated with 50% methanol.

Column dimensions: Length: 22.2 cm; Radius: 1.25 cm

Bed Volume = ðr²h

= 3.14×1.25×1.25×22.2 ml

=109 ml

Mobile Phases used:

50 % Methanol = 650 ml; Flow rate-1 ml/min

70 % Acetone =150 ml; Flow rate- 0.5 ml/ min

Fraction size: 5 ml.

Sample is prepared by dissolving 0.75 g of lyophilized CT powder in 2.5 ml of 50% methanol. 2 ml of this sample is loaded on to the column and the column is eluted with 650 ml of 50% methanol. After elution with 50% methanol, the column is eluted with 70% aqueous acetone. Fractions, of 5 ml volume, are collected in different test tubes. The fractions, having higher range of OD are pooled and again lyophilized.

6.1.3. Estimation of total phenolics

Reagents

1. 1 % HCl in methanol - add 10 mL of HCI to l-L volumetric and bring to volume with methanol.

2. Tannic acid stock solution (0.1 mg/mL) - dissolve 100 mg of tannic acid in 1 L of H_20.

   Prepare fresh immediately before use.

3. Folin-Denis Reagent - add 100 g of$Na_2W0_4 \cdot 2H_20$ to 750 mL of H_20, mix well. Add 20 g of phosphomolybdic acid (correct for waters of hydration if necessary), and mix well. Add 50 mL of H_3P0_4, mix well. Reflux for 2 h (vigorous boil), allow to cool and dilute to 1 L.

4. Saturated sodium carbonate solution - add 700 g of anhydrous Na_2C0_3 to 2 L of H_20.

   Dissolve at 70-80 C, cool overnight. Seed supersaturated solution with crystal of $Na_2COdO\ H_20$, allow to crystalize for 24 h, then filter through glass wool.

Procedure

1. *Preparation of standard curve*: Suitable aliquots of the tannin-containing extract were taken in test tubes and volume made upto 1.0 ml with distilled water. Folin-Ciocalteu reagent (0.5 ml) and 20% sodium carbonate reagent (2.50 ml) were added to the tannin samples.

Treatment	TA sol. (ml)	D.W. (ml)	F-C reagent (ml)	Na_2CO_3 (ml)	TA concn. (mg)	A_{725}
Blank	0.00	1.00	0.50	2.50	0	0
S1	0.02	0.98	0.50	2.50	5	
S2	0.03	0.97	0.50	2.50	10	
S3	0.04	0.96	0.50	2.50	15	
S4	0.05	0.95	0.50	2.50	20	
S5	0.06	0.94	0.50	2.50	25	

2. The contents of the tubes were thoroughly mixed using vortex mixer and kept at room temperature for 40 min.

3. Absorbance was recorded at 725 nm.

Calculations

Absorbance of unknowns is used to predict concentration from curve. These values are multiplied by 50 (because 1 mL of original 50-

mL extract was used). Divide by sample weight, then by dry matter factor to obtain mg ofphenolics/g of dry sample.

6.1.4. Estimation of condensed tannins

Vanillin/HCl procedure for condensed tannins

Reagents

1. 1% HCl in methanol- add 10 mL ofHCl to 990 mL of methanol (use volumetric).
2. Standard solution - add 100 mg of anhydrous catechin to 50-mL volumetric and bring to volume with methanol. Prevent light exposure (use opaque volumetric). If catechin has one water of hydration, use 106.2 mg of catechin. This solution is good approximately 2 wk.
3. 8% HCl in methanol- 80 mL ofHCl brought to 1 L with methanol.
4. 4% vanillin in methanol - 40 g of vanillin (assume anhydrous) in 960 mL methanol. This solution is good approximately 2 wk in opaque container.
5. Vanillin/HCl reagent - mix equal volumes of solutions prepared in Steps 3 and 4. Make fresh *immediately* before use.
6. 4% HCl in methanol- 40 mL ofHCl brought to 1 L with methanol.

Procedure

1. Weigh 0.5-g samples in duplicate.
2. Place in test tubes (15-mL polystyrene tubes with caps).
3. Extract with 10 mL of 1 % (vol/vol) HCl in methanol for 20 min (place tubes in rack and cap, then invert by hand for 20 min).
4. Centrifuge at 1,000 x g for 10 min.
5. Prepare standards (use 10-mL volumetrics). Add 0, 0.5, 1, 1.5, and 2 mL of standard to volumetrics and bring to volume with methanol.
6. Pipette 0.5 to 1 mL of unknowns and standards into test tubes (20=mL disposable culture tubes) in duplicate (two tubes for each extract tube, one serves as blank).
7. To one tube, add 5 mL of vanillin/HCl reagent. To the other tube, add 5 mL of 4% (vol/vol) HCl in methanol (background correction).

8. Place in 30°C H_20 bath
9. After 20 min in H_20 bath, vortex, and read on spectrophotometer at 500 nm.

Calculations

1. Standard curve. Zero serves as blank. The .5, 1, 1.5,2 mL volumes equate to .1, .2, .3, and .4 mg/mL of catechin equivalents (stock has 2 mg/mL, and standards are diluted to 10 mL). Absorbance from these serve as X and .1, .2, .3, .4, and .8 serve as Y in standard curve background correction for standards usually is not necessary.
2. Unknowns. Absorbance of unknown minus absorbance of unknown blank is entered into standard curve to predict concentration. Divide result by sample weight and correct for dry matter factor. Results are expressed as catechin equivalents (mg of CE/g of dry sample).

Ferric Ammonium sulphate method for condensed tannins

Reagents

1. Butanol-HCl reagent (butanol-HCl 95:5 v/v): Mixed 950 ml of n-butanol and 50 ml concentrate HCl (37%).
2. Ferric reagent (2% ferric ammonium sulphate in 2 N HCl): Dissolved 2 g ferric ammonium sulphate in 100 ml of 2 N HCl (16.6 ml of conc. HCl is made to a final volume of 100 ml with water) and stored in an amber coloured bottle.

Procedure

1. Tannin extract (0.5 ml) is taken in glass test tube
2. Butanol-HCl reagent (3.0 ml) and ferric reagent (0.1 ml) are added to it.
3. The contents of the tubes are thoroughly mixed using a vortex mixer.
4. Mouth of the tubes needs to be covered with glass marbles while putting them in a heating block (98 ± 2° C) for 60 min.
5. Tubes are cooled and absorbance recorded at 550 nm against a blank which comprised of an unheated mixture of the above reagents.

Calculation

Condensed tannins (% DM) as leucocyanidin equivalent was calculated by the formula:

$$\frac{A_{550\,nm} \times 78.26 \times \text{dilution factor}}{\%\,DM}$$

The dilution factor was the additional dilution if the absorbance exceeded 0.6. When the extract was from 200 mg sample in 10 ml solvent, the dilution factor was 1.0.

6.1.5. Analysis of non-tannin phenolics

Procedure

1. Hundred mg of commercially available insoluble polyvinyl-polypyrrolidone (PVPP), which binds tannins, was weighed into 100 × 12 mm test tube.
2. Add 1ml distilled water and then 1 ml tannin containing extract and vortexed.
3. The tubes are kept at 4^0 C for 15 min, vortexed again, then centrifuged (3000 rpm for 10 min.) and supernatant was collected.

Calculation

This supernatant has only simple phenolics other than tannins (tannins would go along with PVPP). For measuring phenolic content of the supernatant, the procedure used for the estimation of total phenolics is followed.

6.1.6. Problems in the determination of tannins and possible solutions

Most of the methods available for determining tannins measure tannins relative to one or the other standard, tannic acid (TA) being the most widely used. Use of TA from different commercial sources as a standard influence the tannin levels to a large extent. It is suggested that the commercial source of TA used and the relationship for convertion of absorbance to TA level should be mentioned, and the values obtained by different laboratories should be compared with caution. Vanillin-HCl method is widely used for the determination of condensed tannins (CT). Tannins are generally extracted using aqueous acetone. However, acetone reacted with acidified vanillin to produce chromogen with max at 548 nm which produced a substantial error in the determination of CT. In presence of acetone, the time course of the reaction for catechin and tannins is different depending on the acetone concentration and the reaction temperature. The PPAs take into account only the slope of the 'BSA precipitated per unit TA or per unit feed sample' linear regression and not the y-intercept. It is observed that the y-intercepts of the linear regression for TA and for several feeds especially leaves of trees and shrubs are significantly different from zero. This calls for a change in the present approach (addition of different quantities of tannins to a fixed amount of the protein). The addition of a different quantity of protein to a fixed amount of tannin/tannin-containing extract and determination of the maximal protein bound by the tannin could be a solution to this problem (Makkar and Becker, 1994).

References

Makkar, H.P.S. 1994. Quantification of tannins - A laboratory mannual. ICFARDA, Aleppo, Syria.

Makkar, H.P.S. and Becker K. 1994. Some problems in the determination of tannins and possible solutions. ISHS Acta Horticulturae 381, 782-788.

Foo, L.Y., McNabb, W.C., Waghorn, G.C. and Ulyatt, M.J. 1997. Proathocyanidin from *Lotus pedunculatus*. Phytochemistry. 45: 1689-1696.

6.2. Tannin Bio-assay

6.2.1. Protein precipitation capacity (PPC)

1. 400 mg finely ground selected feed sample is taken in a 250 ml conical flask
2. Prior to extraction of tannins, the pigments and fats from the sample are removed by 40 ml of diethyl ether containing 1% acetic acid. This mixture is shaken thoroughly and kept for 5 min. Diethyl ether is then decanted. After removal of diethyl ether, 20 ml of 70% methanol (Acetone is not used since it interferes with estimation) is added to the defatted and depigmented residue.
3. The flask is shaken for 2 h in an orbital shaker (30 °C, 130 rpm).
4. The contents are then filtered through whatmann filter and different aliquots i.e. 20 µl, 40 µl, 60 µl, 80 µl of filtrate are taken in test tubes, volume is made upto 100 µl with 70% methanol

5. Then, add 1 ml BSA dye (2mg/ml BSA in acetate buffer- 0.2 M acetic acid, 0.17 M NaCl, pH4.9, i.e. 5.7 ml glacial acetic acid, 400 ml distilled water, pH adjusted to 4.8- 4.9 with 4 N sodium hydroxide-16g in 100 ml water, volume it to 500 ml and add 4.93 g NaCl),
6. The contents of the tubes are thoroughly mixed and kept at 4 °C overnight.
7. The following day contents of tubes are centrifuged at 10000 g for 10 minutes.
8. Supernatant is carefully removed without disturbing the precipitate (Pellets may be washed once with acetate buffer-aqueous methanol mixture (2:1, v/v) and used for protein determination.
9. Add 1.5 ml of 1% SDS solution (sodium deodecyl sulfate 1g in 100 ml dist water) and vortwex to dissolve
10. Take 1 ml of the dissolve, add 3 ml SDS-Triethanol amine soln (1% SDS (w/v) and 7% (v/v) triethanolamine in distilled water), then add 1 ml ferric chloride reagent *(0.01 M ferric chloride in 0.1 M HCl- 0.162 g ferric chloride in 0.1 M HCl)*
11. After 15-20 min at room temp, absorbance is read at 510 nm.
12. The amount of protein precipitated can be calculated from the standard BSA calibration curve.
13. A linear regression between BSA precipitated and mg leaf in the aliquot taken for assay is drawn. The slope of the curve (mg BSA precipitated/ mg leaf) represented the protein precipitation capacity.

6.2.2. PEG binding capacity (PBC):

1. 500 mg sample is taken in 50 ml centrifuge tubes, to which 10 ml of PEG solution (8 mg/ml) is added.
2. The tubes are kept in shaker (30°C, 120 rpm) for 21 h and then centrifuged at 17000 rpm for 30 min. followed by collection of supernatants.
3. The supernatants are diluted five times and 0.3 ml of diluted supernatant is deproteinised by sequential addition of 2.1 ml of 1.2% $BaCl_2$, 0.3 ml of 0.3N $Ba(OH)_2$ and 5% $ZnSO_4$.

4. These contents are centrifuged at 2000g (5000 rpm) for 15 minutes.

5. From the centrifuged contents, 1 ml of supernatant is taken, to which sequentially added 1 ml distilled water and 2 ml of 30% TCA+5.9% $BaCl_2.2H_2O$ (needs to be filtered through sintered glass or G1 filter).

6. Protein precipitation is hastened by centrifugation

7. Absorbance is measured at 360 nm after 40 minutes of TCA reagent addition. Blank contains water and TCA/$BaCl_2$ reagent.

8. For standard, stock PEG solution (300 μg/ml) is prepared, which is then diluted to give working concentrations of 30, 60, 90, 120, 150, 180, 210, 240, 270 and 300 μg/ml.

9. The standard curve is constructed by regressing absorbance of standard solutions against log of PEG concentration (say X).

10. The PEG concentrations in experimental samples were estimated from the standard curve by taking antilog of the predicted X.

6.2.3. Radial diffusion assay

Preparation of plates

1. Weigh 2.5 g agarose into 250 ml of the 0.05M acetate buffer (2.85 ml glacial

2. acetic acid, 10.6 mg ascorbic acid in 1L, pH 5.0), heat/microwave to dissolve.

3. Cool down to say 45°C & add 250 mg BSA and dissolve it in the agarose solution (sufficient for preparation of approx. 20 Petri dishes, internal diameter 8.5 cm).

4. Dispense 10-12 ml of agarose-BSA solution and allow it to solidify

5. Cover the Petri dishes and seal them with a strip of Parafilm in order to prevent drying and cracking of the agarose layer (can be stored for 2-3 wks in a refrigerator).

Assay procedure

1. Using a well puncher (4 mm) punch 3-4 wells, far apart, in the solidified agarose in Petri dishes and remove the plugs of agarose from the Petri dishes for making a tiny well

2. Pipette 10-30 ?l (depends on tannin content) of the extract in each well and seal again using Parafilm.
3. Place the plates on a horizontal platform in an incubator adjusted at 30°C.
4. After 96 h, remove the Petri dishes, uncover and measure the diameter of the ring formed.
5. The amount of tannins is proportional to the square of the diameter and the square of the diameter of the ring formed can be calibrated by taking different concentration of tannic acid or any other standard tannin, and the results can be expressed equivalent to a standard.

6.3. Estimation of Saponins

Apparatus and glasswares required

Ether extraction plant, Vacuum oven, Vacuum pump, Analytical balance, Separating funnel, Conical flask, Beakers.

Chemicals

1. Methyl alcohol (AR)
2. N-Butyl alcohol (AR)
3. Solvent ether (Redistilled)

Procedure

1. A known amount of finely ground feed sample (2 - 5g) is extracted for 10 hours in the Soxhlet apparatus with methyl alcohol.
2. The methyl alcohol extract is concentrated to about 10 ml over heating mantle and later it is diluted with 20ml distilled water.
3. This is transferred to a separating funnel (250 ml). To ensure quantitative transfer three washings (5, 3 and 2ml of distilled water) are also given and added to separating funnel.
4. About 10 ml nbutyl alcohol is added and shaken for a minute and the layer containing n-butyl alcohol extract is collected in a 50 ml beaker and the same process is repeated twice using 5 ml portions of n-butyl alcohol.

5. The n-butyl alcohol soluble fraction is concentrated under reduced pressure in three volumes of solvent ether while stirring with the help of glass rod.
6. A white precipitate of saponins is obtained and filtered through pre-weighed whatman filter paper no. 1.
7. Filtration is done immediately as otherwise the precipitate is found to stick on the wall of beaker on keeping. The beaker containing solution of saponins is washed thrice with methyl alcohol (lml portions) to ensure quantitative precipitation.
8. The beaker containing the precipitate is also washed thrice with solvent ether (lml portions).
9. Finally, the filter paper having saponin precipitate is dried under vaccum and weighed.
10. The weight of saponin is expressed on the basis of dry weight of the sample taken.

Calculations

Total saponins (%) = w2 - w1 x 100

Where

w = Dried weight of sample

w1= Dried weight of filter paper

w2 = Dried weight of filter paper alongwith precipitate of saponins

Reference

Yosioka et al. (1974). Tetrahedron, 30: 707.

6.3.1. Extraction and partial purification of saponins from *Asparagus adscendens* root

1. *Asparagus adscendens* root powder was extracted with petroleum ether for 6h.
2. Lipid extracted powder was kept in an oven at 39° C for 12 hours in order to evaporate residual petroleum ether.
3. Fat free dried root powder (20 g) was extracted with methanol (100 ml) using shaker set at 30° C and 120 rpm speed for 8 hours.
4. The contents were squeezed through 2 layers of muslin cloth and filtered using whatmann No.1 filter paper. The above process was repeated with 5*20ml methanol and the filtrates were pooled.

5. The methanol in the filtrate was evaporated under reduced pressure to obtain viscous mass. (dark red in colour)
6. The concentrated extract was dissolved in 15 ml distilled water and partitioned with n-butanol (2*15ml; 1*10 ml)
7. The butanol extract was concentrated under reduced pressure at 60°C
8. The butanol extracted solids (817 mg) were dissolved in 5 ml methanol
9. Saponins in the extract was precipitated by adding 15 ml diethyl ether and filtered through Whatmann No.1 filter paper. The process was repeated with 2*1 ml methanol and 2*3 ml diethyl ether for complete recovery.
10. The white precipitate (saponins) on filter paper (437 mg) was recovered with 2*3 ml methanol..
11. Saponins in the methanol was precipitated with 18 ml acetone and the same process was repeated with 2*1 ml methanol and 2*3 ml acetone. The white precipitate was recovered by filtration using whatman No.1 filter paper.
12. The white precipitate was suspended in 10 ml methanol water (1:1) and the methanol and the traces of diethyl ether was evaporated. The aqueous portion was freeze dried to obtain white saponin crystals (50 mg).

6.4. Estimation of Gossypol in Cotton Seed Meal

Cotton *(Gossypium)* seed meal (CSM), the by-product of cotton seed oil industry is widely used as an animal feed. However, its utilisation as an animal feed is limited due to the presence of gossypol, a polyphenolic binaphthyl aldehyde, which is toxic to non-ruminants such as poultry and swine and young ruminants such as calves and lambs. The gossypol (free gossypol, FG) present primarily in the pigment glands of cotton seed is mostly converted to bound gossypol due to condensation of aldehyde group with free amino groups of proteins to form a *schiff* base. Formation of *schifl* base, though results in the detoxification of gossypol to some extent, it also lowers the nutritive value by reducing the availability of lysine, the limiting amino acid of cotton seed. Thus, bound gossypol, consequently total gossypol (TG) (free plus bound) content is an important factor of cotton seed meal.

Principle

The method for total gossypol estimation involves oxalic acid hydrolysis of the bound gossypol in a methyl ethyl ketone-water azeotrope, partitioning the liberated gossypol into chloroform and quantification by second derivative UV spectrophotometry at 300nm, which permits direct quantification of all compounds containing naphthalene nuclei, chromogenic reaction is not required (SD, 4.0; overall recovery, 89.5%). The method of free gossypol estimation involves extraction of free gossypol with aqueous acetone hydrolysis of the soluble bound forms of gossypol with hydrochloric acid, partitioning of the pure compound into chloroform and analysis by derivative UV spectrophotometry at 300nm, the reaction requires no chromogenic reagent (SD, 4.0%; overall accuracy, 91.2%).

Apparatus

Double beam UV-visible spectrophotometer with 10 mm quartz absorption cells, Water bath, Rotary vacuum evaporator, Laboratory mill equipped with 1 mm screen, Wrist action mechanical shaker

Reagents and chemicals

1. Aqueous acetone: Mix 700 ml acetone with 300 ml glass distilled water.
2. Methyl ethyl ketone-water-azeotrope:
3. Mix 1106 ml reagent grade methyl ethyl ketone with 110 ml water and distill, rejecting the first 100 ml of distillate. Azeotrope distills at 73.5°C. Store in brown bottle.
4. Oxalic acid solution(O.l M):
5. Dissolve 12.6g oxalic acid dihydrate in 1000 ml methyl ethyl ketonewater-azeotrope and store in brown bottle.
6. Barrurn acetate solution-0.5 M:
7. Dissolve 136.73g barium acetate hydrate in 1000mi distilled water.
8. Solvent mixture:
9. Mix 715 ml ethanol, 285 ml distilled water, 200ml peroxide free diethyl ether and 2 ml reagent grade glacial acetic acid. Store in brown bottle.

10. Other chemicals: Chloroform, sodium sulphate and hydrochloric acid.
11. Gossypol stock solution:
12. Weigh accurately 12 mg of standard gossypol acetic acid (89.62% gossypol by weight; Sigma Chemicals) and dissolve in 50 ml solvent mixture. Aliquots of stock solution are further diluted to give working solutions in the range of 0.7-4 ug/rnl. Stock and working solution are prepared daily and protected from light throughout the analysis.

Procedure

1. Standard curve for total and free gossypol

Measure optical density (OD) of working solutions at 300 nm. Plot calibration curve taking concentration of gossypoluig/ml) on X-axis against corresponding O.D. on Y-axis. Regression of gossypol (ìg/ml) concentration (y) on O.D. (x) revealed a linear response within the observed range with a correlation co-efficient (r) of 0.999 giving a model equation:

$$y = 0.006315 + 39.7326\ x$$

Where, x = O.D. and y = gossypol concentration (μg/ml)

2. Sample preparation

Grind approximately 50 g cotton seed meal in laboratory mill through 1 mm screen. Care should be taken to avoid overheating and oil expression while grinding, especially when seeds are ground.

Estimation of total gossypol

1. Accurately weigh 1 g of ground CSM into 100 ml volumetric flask and add 25ml oxalic acid solution.
2. Place flask in 75°C water bath, allow to equilibrate, then stopper flask and heat for 6 hours.
3. Remove flask from water bath, cool to room temperature and add 25ml aqueous acetone followed by 5 ml barium acetate solution.
4. Mix contents of flask and dilute to volume with aqueous acetone. Let it stand 10 min. for complete precipitation of formed barium oxalate

5. Filter through Whatman No. 41 filter paper into glass stopper flask, discarding first 10 ml of filtrate.
6. Pipet 25ml aliquot of filtered extract into 250 ml separatory funnel and add 50 ml chloroform followed by 100 ml distilled water acidified with 1 ml hydrochloric acid.
7. Shake funnel 3 min to accomplish partitioning, let stand 5 min. and filter lower organic layer through anhydrous sodium sulphate on Whatman No. 40 filter paper into 100 ml glass stoppered round bottom flask.
8. Rinse sodium sulphate by filtering through filter paper with 10 ml chloroform and evaporate at 40°C.
9. Dissolve dried residue with 25ml solvent mixture and stopper flask.
10. Pipet appropriate sample solution into 10 ml volumetric flask and dilute to volume with solvent mixture. Size of aliquot depends on total gossypol content of sample:

    For sample with expected total gossypol content <3000ppm, use >1.0 ml aliquot; 3000 - 5000 ppm, 1 ml aliquot; 5000 7000ppm, 0.7 ml aliquot; 7000 - 9000 ppm, 0.6 ml aliquot; 9000-11000 ppm, 0.5 ml aliquot and 11000 - 13000 ppm, 0.4 ml aliquot. For maximum precision, resulting final solution should contain about 2.5 μg gossypol/ml.
11. Read O.D. of final solution against solvent mixture at 300 nm.

Calculations

Calculate total gossypol content of CSM as follows:

Total gossypol (ppm) = (2000 × C)/(V × W)

Where,

C = Gossypol concentration (ìg/ml) of final solution from standard curve,

V = Volume of aliquot taken (ml),

W= Weight of sample (g)

Estimation of free gossypol

Estimation of free gossypol involves extraction, hydrolysis and clean up procedures.

1. Extraction

a) Transfer an accurately weighed sample of ground (lmm screen) cotton seed meal (3g) to a 250 ml glass stoppered Erlenmeyer flask after covering the bottom of flask with glass beads (4 mm diameter).

b) Add 100 ml of 70% aqueous acetone and shake the flask for 1 hour on a wrist action mechanical shaker.

c) Filter the contents of flask through Whatman No. 40 filter paper into another flask, after discarding first 5 ml of filtrate.

2. Hydrolysis and clean up

i) Pipet 2 ml of aliquot of filtrate into a 50 ml volumetric flask and add 1 drop (0.5 ml) of hydrochloric acid.

ii) Heat the flask contents on a water bath at 65°C for 1 hour and cool it to room temperature.

iii) Transfer the contents to a 250 ml separatory funnel with two rinses, 3 ml each, of 70% aqueous acetone. Add 100 ml of acidified (1 ml hydrochloric acid) distilled water into the funnel.

iv) Extract the formed suspension with 50 ml of chloroform and filter lower organic layer through anhydrous sodium sulphate on Whatman No 40 filter paper into a 100 ml glass stoppered round bottom flask.

v) Rinse sodium sulphate and filter paper with additional 5 ml of chloroform.

vi) Evaporate the filtrate to dryness in a vacuum oven at 35°C.

vii) Dissolve the dried residue with 25 ml solvent mixture and store in the flask until analysed.

N.B. Values of gossypol are unchanged on standing up to 6 hours.

viii) Read O.D. of sample solution against solvent mixture at 300nm.

Calculation

The concentration of free gossypol in the sample is calculated by referring to calibration curve after multiplying by appropriate dilution factor as shown above.

$$\text{Free gossypol (ppm)} = \frac{100 \times C}{W}$$

Where,

C = gossypol concentration (ìg/ml) of final solution from standard curve

W= weight of sample (g)

Estimation of bound gossypol

Estimate bound gossypol by subtracting free gossypol from total gossypol.

Bound gosssypol (ppm) = Total gossypol (ppm) - Free gossypol (ppm)

References

Botsoglou, N.A., 1991. Determination of 'free' gossypol in cottonseed an dcottonseed meals by Secdond-Derivative Ultraviolet spectrophotometry. J. Agric. Food Chern., 39: 478 - 482.

Botsoglou, N.A. and Kufidis, DC, 1990. Determination -of total gossypol in cottonseed and cottonseed meals by Derivative UV spectrophotometry. J. Assoc. Off. Anal. Chern., 73: 447 - 451.

6.5. Estimation of Mimosine

Principle

Mimosine is extractable in dilute HCl and it gives colour reaction with FeCh reagent which can be measured at 535nm.

Apparatus and glassware

Spectrophotometer, Vortex mixer, Hot plate, Whatman No.1 filter paper, Nylon cloth, Beaker-250ml, Measuring cylinder-100ml, Funnel, Conical flask-100ml, Test tubes

Reagents

1. Hydrochloric acid (0.1 N).
2. Ferric chloride solution (0.5%) in O.lN HCl 3.8tandard mimosine solution (lmg/ml) in 0.1 N HCl

Procedure

1. Preparation of standard curve

Take 0, 0.1, 0.2, 0.4, 0.6, 0.8 and 1ml of standard mimosine solution (lmg/ml) and proceed as follows:

Tube No	Std Mimosine	Vol of 0.1N HCl	$FeCl_3$ sol	Mimosine(mg)	A_{535nm}
Blank	0.0	9.6	0.4	0.0	
1	0.1	9.5	0.4	0.1	
2	0.2	9.4	0.4	0.2	
3	0.4	9.2	0.4	0.4	
4	0.6	9.0	0.4	0.6	
5	0.8	8.8	0.4	0.8	
6	1.0	8.6	0.4	1.0	

Take OD at 535 nm and draw the graph against the concentration of mimosine.

2. Extraction of the sample

a) Take 1-1.5g dried and ground sample it) a beaker. Add 30-40 ml of O.IN HCI, stir with glass rod and boil gently for 5 minutes on hot plate.

b) Cool and allow the material to settle down.

c) Filter through a nylon cloth in a 100mi measuring cylinder.

d) Again boil the residue with 20-25 ml of O.IN HCI, cool and filter. Repeat the procedure one more time.

e) Make up the volume of filterate to 100 ml with O.IN HCl.

3. Decolourisation of extract

Take 10 ml filterate in a conical flask. Add 10 ml of O.INHCI and a pinch of activated charcoal. Shake to mix thoroughly and heat for 5 minutes. Cool, allow to settle and filter through Whatman filter paper No.1 or 2. If the filterate is not yet clear, again treat it with charcoal and collect the clear filtrate.

4. Estimation

a) Blank - Take 1 ml clear filtrate and make it 10 ml with O.IN HCI and take its *G.D.* at 535 nm.

b) Sampel- To 1ml filtrate add 8.6ml O.lN HCI and 0.4 ml ferric chloride solution and read it at 535 nm.

Observations

Tube No.	Sample	Volume of sample(ml)	O.lNHCI (ml)	Ferric reagent (ml)	A_{535nm}
1	Blank	1.0	9.0	-	
2	Sample 1	1.0	8.6	0.4	
3	Sample 1	1.0	8.6	0.4	

Calculation

Corrected OD of sample = OD of sample - blank. Compre it with standard curve and express the mimosine content in sample on DM basis.

6.6. Estimation of Total Glucosinolates

Principle

The glucosinolates (GSL) of feed sample are dissolved in water. 0.1 N AgNOa and 95% ethanol is mixed with the solution. The solution with ferric ammonium sulphate in acidic medium is titrated against *N/100* potassium thiocynate.

Apparatus

Grinder, heater, filter paper, buchner funnel, wash bottle, beakers, burette.

Reagents

(i) O.lN AgNOa
(ii) 95% Ethanol
(iii) 8% *w/v* Ferric ammonium sulphate
(iv) 6N-HCl
(v) 0.01 N Potassium thiocynate solution

Procedure

1. 10g of ground sample is enzyme deactivated by dropping it into 250 ml of boiling water for 5 minutes. The enzyme deactivation step is normally not allowed to prolong beyond 5 min. to avoid the breakdown of glucosinolates.

2. The contents are filtered on a Buchner funnel.
3. The residue is washed with 50ml hot water three times and filtered in the same manner and the volume of total filtrate is made up to 500ml.
4. An aliquot of 25 ml is taken in a beaker to which 10 ml of 0.1N $AgNO_3$ and 25 ml ethanol (95%) are added.
5. The contents are refluxed on a water bath for 45 min. and then cooled to room temperature
6. Volume is made to 100 ml with distilled water and filtered through ordinary filter paper.
7. An aliquot of 25 ml of supernatant is taken in a 125 ml flask containing 2 ml of 6N HNO_3 and 6 ml of 8% w/v ferric ammonium sulphate solution.
8. The homogenous mixture is titrated against 0.01N potassium thiocynate till a pale salmon colour is obtained. A blank is also run with each determination.

Calculation

$$\%\,\text{Glu cos inolates} = \frac{(\text{Blank} - \text{tiration}) \times 4 \times 0.01 \times \text{Mole wt. of GSL} \times \text{Total volume}}{1000 \times 2 \times \text{Sample wt.} \times 25}$$

Total volume = 500 ml; Mol wt. of GSL = 411

Reference

McGhee, J.E., Kirk, L.D. and Mustakas, G.G. 1965. Total glucosinolates method for *Brassica* species. J. Assoc. Off. Anal. Chern. 42: 889-891.

6.7. Estimation of Cyanogenic Glycosides

Qualitative Test

Apparatus

Stoppered bottle, Filter paper strips

Reagents

i) Picric acid solution (1%)

ii) Sodium carborate (Na2CO3) (10%)

Procedure

1. Filter paper strips are dipped in 1% picric acid solution and after drying, they are further dipped into 10% Na2C03 solution and dried again. These strips are stored in stoppered bottle.
2. Finely chopped small amount of plant material is placed in test tube. Now insert a piece of moistened sodium picrate paper in tube while taking care that it does not come in contact with sample. Add few drops of chloroform and stopper tube hermatically. The sodium picrate paper gradually turns orange, then brick red if plant tissue contains cyanogenic glycosides. The rapidity of change in colour depends upon amount of free HCN present.
3. This test works well with fresh plant materials, but relatively dry substances particularly seeds of various plants should be ground and moistened with H20 and allowed to hydrolyse in stoppered test tube containing sodium picrate paper.

6.7.1. Titrimetric method for quantitative test

6.7.1.1. Acid titration method

Apparatus and glass wares

Micro kjeldahl distillation apparatus, Kjeldahl flasks, Conical flasks, Burette, Pipette, Gooch crucible.

Reagents

(i) Silver nitrate (0.01 N)

(ii) Nitric acid concentrated (Potassium cynide (KHCN) - (0.02N)

(iii) Ferric alum indicator.

Procedure

1. About 10-20 g finely ground sample (sieve no. 20) is placed in 800 ml kjeldahl flask.
2. Add 100 ml H_20 and macerate at room temperature for 2 h.
3. In this flask further add 100 ml H_20 and steam distill for collecting distillate in 20 ml 0.01N $AgNO_3$ acidified with 1 ml HNO_3.
4. Before distillation adjust apparatus so that tip of condenser dips below surface of liquid in receiver.

5. When 150 ml distillate is collected then it is passed through Gooch crucible.
6. Wash receiver and Gooch with little H_20 and titrate excess $AgN0_3$ in combined filtrate, and washings with 0.02N KHCN, using ferric alum indicator.

Calculations

1 ml 0.02 N AgN03 = 0.54 mg HCN

6.7.1.2. Alkaline titration method

Apparatus

Kjeldahl flask, Conical flask, Distillation apparatus (micro), Burette, Pipette.

Reagents

i) Sodium hydroxide (0.5 g in 20 ml H20)
ii) Ammonium hydroxide (6N)
iii) Potassium iodide (50/0)
iv) Silver nitrate (0.02N)

Procedure

1. About 10-20g of finely ground (seive No. 20) sample is placed in 800 ml kjeldahl flask and add 200 ml H20 and let stand for 2 - 4 hrs (analysis is done with apparatus completely connected for distillation). Now steam distill and collect 150-160 ml distillate in NaOH solution (0.5g in 20 ml H20), and dilute to a definite volume.
2. To 100 ml distillate (preferable to dilute 250 ml and titrate 100 ml aliquot) add 8 ml 6N NH40H and 2 ml 5% potassium iodide solution and titrate with 0.02 N AgN03 using micro burette. The end point is faint but permanent turbidity and may be easily recognised, espicially against black background.

Calculations

1ml 0.02N $AgNO_3$ = 1.08 mg HCN (1 Ag equivalent to 2CN)

Reference

AOAC (1995). In Natural Toxins (49th chapt.)", pp. 37-49. Official Methods of Analysis of AOAC International, 16th Edn., (Vol. II) , Virginia, USA.

6.8. Qualitative Test to Detect Ricin

Principle

The Ricin present in saline extract of castor *(Ricinus communis)* bean meal (CBM) exerts haemagglutinating activity and can be utilized for detecting the presence of CBM in feeds.

Reagents

i) Normal saline (0.9%, NaCl)

ii) EDTA

iii) Solution containing 0.14 M Sodium chloride, 20 mM Sodium phosphate per 100 Ill (pH 7.1)

iv) Bovine serum albumin

Procedure

1. Preparation of crude extract

a) Stirr 50 g of test feed mechanically in 150 ml normal saline (0.9%, NaCl) for 1 h.

b) Filter the contents through muslin cloth and centrifuge the filtrate at 2000 rpm for 5 minutes.

c) An aliquot of clear supernatant is further tested for haemagglutinating activity of ricin by comparative qualitative test using plate agglutination technique.

2. Plate agglutination test

1. Collect 10 ml of blood from chicken's wing vein in equal volume of normal saline having EDTA @ 1mg/ml blood and centrifuge at 2500 rpm for 20 minutes to sediment red blood cells (RBC).
2. Wash RBC thrice with normal saline through centrifugation and dilute further with normal saline to a final suspension of one percent.
3. Carryout micro-haemagglutination test in Lambro plate as per the method of Olsnes *et al.* (1974) with slight modification as described below.
4. To each well add 100 µl of a solution containing 0.14 M sodium chloride, 20 mM sodium phosphate (pH 7.1) and 100 ìg bovine serum albumin per ml. add 100 ìl of the crude extract to the first well and make serial dilutions by transferring after thorough mixing @ 100 µl to the next well and so on.

5. Subsequantly, add 25 µl of normal saline washed 1% erythrocytes to each well and mix gently.

6. Incubate the plate at 37°C and take the reading after 10 minutes. In case of agglutination, the erythrocytes will be sticky and cover the bottom of the well as a thin film or matrix (+), whereas, nonagglutinated RBC settle as button (-) at bottom in the centre of the well.

7. Express haemagglutinating activity as HA unit (the reciprocal of the end point dilution).

Reference

Olsnes, S., Saltvedt, E. and Alexander, P. 1974. Isolation and composition of galactose binding lectins from Abrus precatorius and Ricinus communis. J. Bioi. Chern,. 249: 803.

6.9. Detection of Argemone

Principle

Argemone m acidic medium gives orange-red colour with ferric chloride.

Equipments

Extraction apparatus, filter paper, water bath or hot plate, test tubes, pipettes.

Reagents

i) Petroleum ether (40-60°C)

ii) Cone. HCl

iii) *Ferric chloride solution*: Dissolve 10 g of ferric chloride crystals in 10 ml cone. HCl and add 90 ml water to it. Filter if required.

Procedure

1. Extract 100 g feed/plant material with petroleum ether.

2. Decant the ether extract through filter papers.

3. Evaporate ether, cool and transfer to test tube.

4. Add 3 ml HCl and immerse in boiling water bath for 2 minutes and mix thoroughly by shaking.

5. Repeat twice and then keep the tube until acid and fat layers have separated completely.
6. Pipette out fat layer as much as possible and slowly add 1 ml of ferric chloride along the side of tube taking care to prevent mixing of fat and acid layers.
7. Immerse the tubes in water bath and allow to remain for 10 min. An orange-red precipitate in acid layer indicate the presence of argemone.

Calculation

By making standards of argemone in various concentrations, the unknown quantity of argemone in feed samples can be measured by calorimetric method.

Reference

IS 1932 - 1961: Specifications for mustard and rape seed oil cake as livestock feed. Bureau of Indian Standards, Manak Bhavan, New Delhi.

6.10. Estimation of Nitrates and Nitrites

Principle

Nitrate is reduced to nitrite by zinc and manganese sulphate. The reaction then consists of diazotisation of sulfonic acid by the nitrite ion and subsequent coupling with l-naphthylamine to form a red dye. Copper aids the reaction at low ppm level (0.2) but over 1.2 ppm it interferes. Interfering iron is complexed with citrate. Reaction is optimum at pH range 1. 7 to 3.0.

Reagents and chemicals

1. 20% acetic acid reagent:

   200 ml acetic acid + 5 ml copper sulphate (0.1572g CUS04. 5H20/ l H20, w/v)

   Diluted to 11 with distilled water.

2. Bray's indicator:

   (a) lOOg barium sulphate (BaS04)

   (b) 109 manganese sulphate (MnS04. H20) (c) 2 g zinc (metallic Zn)

(d) 75 g citric acid

(e) 4 g sulfanilic acid

(f) 2 g I-naphthylamine

Zinc and citric acid are powdered, manganese sulphate, zinc, sulfanilic acid and 1-naphthylamine are mixed separately with a part of barium sulphate and then mixed with citric acid and remaining part of barium sulphate. The indicator is stored in blackened bottle away from light.

3. *Hydrochloric acid (O.IN)* : 8.18 ml. 37 - 38% Helll H20.
4. Activated charcoal.
5. *Standard nitrate solution*:

   Stock solution [Nitrate ($N03^{2}$-)= 1mg/ml]: 1.37g sodium nitrate in 1 litre distilled water

   Working solution (10 ug/rnl): Dilute 1 ml of stock solution to 100 ml with distilled water.

Procedure

A. Sampling of feed

At any sampling time separate samples from different places in the field, stack or silo are to be obtained and each is to be analysed separately. Moist sample (silage) should be kept frozen until analysed. Green may be dried immediately in an oven at 60°C.

B. Analysis

1. 1g air dry material or 3-5 g moist sample (nitrate content ranging between 1 to 8 mg) is taken in a conical flask (150ml).
2. 100ml of 0.1N HCI is mixed and the content is shaken occasionally to stand for 1h.
3. If the extract is strongly coloured it is to be decolourised with activated charcoal.
4. The content is then filtered through Whatman filter paper (No.1)
5. The sample extract, blank and standard are run according to the protocol given below:

Protocol

Reagents/Chemicals	Sample	Blank	Standard			
			1	2	3	4
Extract (ml)	1.0	0.0	0.0	0.0	0.0	0.0
Distilled wter (ml)	0.0	1.0	0.0	0.25	0.5	0.75
Working standard (ml)	0.0	0.0	1.0	0.75	0.5	0.25
Conc. of standard (μg)	-	-	10.0	7.5	5.0	2.5
Reagent (1) (ml)	9.0	9.0	9.0	9.0	9.0	9.0
Bray's Indicator (g)	0.5	0.5	0.5	0.5	0.5	0.5

6. Shake well and keep away from strong light.
7. Centrifuge at 3000 rpm for 5 min. Remove any film on top and take the clear red supernatant.
8. Read at 520 nm and find out concentration of nitrate in sample from the standard curve drawn taking OD against concentration.

   N.B. For nitrite, the procedure follows as usual by omitting Zn and $MnSO_4H_2O$ from Bray's indicator.

Calculation

$$\text{Total nitrate (nitrate + nitrite)} = \frac{\text{Conc. of nitrate } (\mu g/ml) \text{ in final solution from standard curve}}{\text{g sample taken (DM basis)} \times 10}$$

$$\text{Nitrite concentration } (\mu g/ml.) = \frac{\text{Conc. in final solution .A obtained from standard curve}}{\text{g sample taken (DM basis)} \times 10}$$

Reference

Wiseman, H.G. and Jacobson, W.C. (1965). J. Agri. Food Chem., 13: 36 - 39.

6.11. Estimation of Oxalic Acid

Oxalates are present in free and salt forms in the feeds and fodders. They are rich in beet and spinach and among straws/stovers, paddy straw contains substantial amounts. Conversely, its low levels are present in peas, beans and brassicas, *etc.* In monogastric animals like, pigs, poultry and rabbits oxalates containing diets reduces growth as well as calcium retention. However, ruminants are able to decompose oxalic acid with the help of rumen microorganisms.

Principle

It is extracted in hydrochloric acid (HCI)and precipitated in the form of calcium oxalate by adding calcium chloride. The precipitate is then washed and titrated with N/20 potassium permanganet (KMn04) in the presence of dilute sulphuric acid (H2S04) at 70°C. One ml of N/20 potassium permanganet is equivalent to 2.25 mg of oxalate. (Abaza *et al.* 1968).

Reagents

i) 6N hydrochloric acid - Add equal amounts of concentrated HCI and distilled water to prepare 6N HCL

ii) Calcium chloride solution (5%) - 5 g of calcium chloride anhydrous (Ca Clz) is dissolved in 100 ml distilled water.

iii) Sulphuric acid (H2S04) in water (1:4 ratio)- A volume of sulphuric acid is slowly poured into 4 volumes of distilled water.

iv) Concentrated ammonia (NH40H)

v) Methyl red indicator.

Procedure

1. To weigh 2g of sample in a 250 ml volumetric flask, add 190 ml of distilled water and 10 ml of 6NHCI and boil the contents on a water . bath. Make up the volume after cooling and filter the supernatant through Whatman (No. 41) filter paper.
2. Pipett out 50 ml of supernatant in a beaker and add 20 ml of 6NHCL Evaporate to about half of its volume and filter. The precipitate is washed in between to make the volume to about 125 rnl.
3. Add 3-4 drops of methyl red to the filtrate followed by concentrated ammonia till the solution turns faint yellow. Heat the contents to 90100°C and filter after (Whatman No. 41 filter) cooling to remove the precipitated impurities, if any. Add 10 ml of 5% CaCh with constant stirring and allowed to stand for 24 hours.
4. FIlter through Whatman No. 41 filter paper, wash the precipitate several times with hot water to make it free from calcium ions. Transfer the precipitate to the original beaker by washing with

distilled water. Then add dilute sulphuric acid solution (1:4) till the precipitate is completely dissolved.

5. Warm the contents to 70°C and titrate with N/20 $KMnO_4$ near to end point and stirr the filter paper to remove to the side contents of the beaker, wash with hot distilled water and complete the titration.

Calculation

Oxalate (g/100g) = N/20 $KMnO_4$ used (ml) × 0.00225 × 250/50 × 100/2

Oxalate (%) = N/20 $KMnO_4$ used (ml) × 0.5625

Reference

Abaza, R.H., Blake, J.T. and Fisher, E.J. (1968). J. Assoc. Off. Anal. Chern., 51: 963.

6.12. Estimation of Typsin Inhibitors

Principle

The casein is used as a substrate for assaying the activity of the trypsin enzyme. The inhibition of the trypsin enzyme activity is measured in the extract, which is obtained from the defatted feed material (Roy and Rao, 1971).

Reagents

i) Trichloro acetic acid (TCA) solution - (5%)

ii) Casein solution in phosphate buffer - (2%)

iii) (0. 1M) Sodium phosphate buffer (pH-7.5)

iv) Trypsin solution - (5mg/ml)

v) Hydrochloric acid - (0.001 M)

Procedure

1. About 4g of the finely ground deffated material feed is taken in stoppered 250 ml conical flask and treated with 40 ml of 0.10 M sodium phosphate buffer (p H 7.5) and 40 ml of distilled water.

2. The suspension in the flask is shaked for 3 h on shaker and centrifuged at 700g for 30 min. at 15°C.

3. The supernatant is diluted in such a way so that 40-60% activity of control enzyme is inhibited.

4. Prepare an incubation mixture containing 0.5 ml trypsin solution, 2.0 ml. casein solution (2%), 1.0 ml O.IM sodium phosphate buffer, 0.4 ml HCl (O.OOIM) and 0.1 ml extract or supernatant (volume of incubation mixture is kept 4.0 ml).
5. This mixture is incubated at 37°C for 20 min. and then 6.0 ml 5% TCA is added to stop the reaction.
6. Simultaneously blanks should also be run.
7. Absorbance is recorded on the Spectrophotometer at 280 nm.

Calculation

One trypsin unit is defined as an increase of 0.01 absorbance unit at 280nm in 20 min. for 10 ml reaction mixture and the trypsin inhibitory activity as number of trypsin units inhibits (TU!) is expressed per mg of protein.

Note

(1) To express the value of specific activities, the total TUI expressed per mg of protein.

(2) The value for trypsin inhibitory activity must have about 40 to 60 percent inhibition to have a reproducible and accurate results.

Reference

Roy, D.N. and Rao, P.S. (1971). J. Agric Food chern., 19: 257-259.

6.13. Estimation of Aflatoxin

Aflatoxins are extremely potent carcinogens to many animals. Neither effect of aflatoxins on man nor possible routes of entry are presently known. So take particular precautions: Use glove box, bleach the accidental spills with 1% NaOCI add 5% aqueous acetone, rinse all glasswares exposed to aflatoxin with methyl alcohol (80%), add NaOCI (17%) soln and after 2hrs add acetone (5%), allow to react for 30 minutes and then wash thoroughly.

Scope

This method is applicable to groundnut, maize, wheat, barley and many other feed samples that does not have colouring substances other than carotenoids.

Principle

Aflatoxin B1 (AF-B1) is extracted from finely ground sample with chloroform. The extract collected is further purified using column chromatorgraphy and then purified extract is condensed by evaporating it to a known volume and subjected to TLC (Thin layer chromatography) for its determination.

In case of TLC, the amount of aflatoxin B1 is estimated visually by comparing fluorescence intensity of aflatoxin in the sample with that of known quantity of AF-B1 standard or more accurately by separating the fluorescent area from TLC plate and dissolving in methanol and comparing the intensity of toxin at the spectrophotometer with different concentrations of standard.

Preparation of sample

Aflatoxins are likely to occur in few pockets at high concentration which may not be randomly distributed. Perform sampling and sample preparation, keeping this point in mind. Due to possibility of pockets of contamination, include total laboratory sample (1-2 Kg) in sample preparation. Aim at maximal particle size reduction and thoroughness of mixing to achieve effective distribution of contaminated portions. While handling large samples, coarse grind and mix entire sample, remove 1I20th and regrind this portion to finer size (1200 microns) for drawing analytical sample. Where divide by riffling is not practical, composite number of small randomly taken portions for analysis. If the lot sample is already in the form of a powder, mix it thoroughly and draw a sample for analysis by quartering it.

Apparatus

Mechanical shaker: 'Wrist Action' or equivalent, fitted to hold 500 ml Erlenmeyer flasks having standard tapered ground glass joints.

Thin layer Chromatography equipment: Standard thin layer chromatography apparatus is consisting of applicator, mounting board, spotting templates, standard size plates (20 x 20 em), plate holder, development tank with glass cover.

Preparation of TLC plates

Plates are coated with silica gel slurry to a thickness of 0.25 to 0.30 mm using standard applicator. Air dry coated plates are activated for 2h at 105°C and store in a desiccating chamber until use.

1. Clean up columns - 300 x 10 urn long with 35 urn porous polyethylene frit, with stopcock.
2. Separating funnel (500 ml), beakers (500 ml & 100 ml), measuring cylinders, funnels, etc.
3. Viewing chamber fitted with UV lamp having long wave (360 nm).

Reagents

All reagents should be of analycal grade *(ARJGR)* supplied by reputed manufacturers.

1. Acetone, Acetonitrile, Benzene, Chloroform, Diethyl ether, Hexane/Petroleum ether (60-80°C) and Methanol.
2. Column wash solvent - Anhydrous diethyl ehter : Hexane (3: 1)
3. Aflatoxin eluting solvent - Chloroform: Methanol (97:3)
4. Standard aflatoxin dissolving solvent - Benzene: acetonitrile (98:2)
5. Anhydrous sodium sulphate (Na_2SO_4)
6. Silica gel for column chromatography (60 - 120 mesh): Dry it in about 1 inch thick layer for 2h at 105°C, cool in a vapour tight container at 1% water (1 ml in 100g) , shake well and equilibrate at least 15h. Store in screw capped bottle in vapour tight container.
7. Silica gel G for TLC (60 mesh)
8. *Aflatoxin standard solution:*

   Prepare a stock solution containing 20 ug/ml of AFBI from high purity crystaline Bi.

   From the stock solution prepare working standards containing 1-5 ug/rnl aflatoxin Bi

   Standard solution should be stored in well stoppered flasks/tubes at 0°C.

Procedure

1. Finely ground feed sample (50g) is taken into an Erlenmeyer flask and mixed thoroughly with 25 ml distilled water with the help of glass rod to prevent lump formation. Add 250 ml of chloroform and shake on a wrist action shaker for 30 minutes. Then filter

the contents through Whatman filter paper no. 1, and collect 50 ml portion of clear filtrate. Condense the 50 ml portion on a water bath to 1-2 ml and save the concentrated portion for column clean up.

2. Prepare silica gel column as follows

   Take a glass column of size 1 x 30 cm equipped with a sintered plate or place a ball of glass wool loosely in the bottom. Clamp it on a stand and add about 20 ml chloroform keeping the stopcock closed. Add in sequence from the bottom to top 1 g of anhydrous sodium sulphate, 2g of silica gel (60 - 120mesh) and finally 1.5 g of anhydrous sodium sulphate. When the three additives have formed proper layer, allow the solvent to run down. As soon as the chloroform crosses the top of the column (Na2S04) load quantitatively the 2 ml concentrated extract of sample and allow it to percolate through the column. As soon as the sample crosses the top of the column bed, give washings to the material in the column with 75 ml of solvent mixture (diethyl ether: hexane, 3: 1) and discard it. When the washing is over, add 60 ml of the eluting solvent mixture prepared using the chloroform: methanol (97:3). Collect this elute in a 100 ml conical flask.

3. Evaporate until dry either on a rotary evaporator or on a water bath and transfer the residue quantitatively to a 5 ml vial. Make up the volume with benzene: acetonitrile or with chloroform to 0.2 ml (200fll). Shake well to dissolve the aflatoxin. This dissolved residue is used for spotting on TLC plate.

4. Prepare thin layer chromatography plates using a slurry of silica gel G (60 mesh) in distilled water. A slurry prepared by mixing 12 g silica gel G in 20 ml of distilled water is optimum to coat two plates of 20 x 20 em size. Plates should be activated before spotting in an oven maintained at 105±2°C for an hour.

5. Spot 5 III or less of the cleaned sample extract with suitable dilutions on to the silica gel plate alongwith 5 III or less of aflatoxin Bi standard solution containing 2, 4 and 6 ng of the toxin per 5 III or less of solvent (Benzene: acetonitrile, 98:2). Develop the plates in the following solvent systems (200ml) independently.

   a) *Chloroform* : Acetone : Methanol (90:8:2)

   b) *Chloroform* : Methanol (95:5)

   c) *Toluene* : Isoamyl alcohol : Methanol (66:32:2)

Allow the plates to develop for 30 - 40 minutes in anyone of the above solvent systems. After this period remove the plates, air dry and visualise the spots under longwave range UV cabinet, comparing the sample fluorescent spots with standard. After identifying the standard spot which just matches with the sample or the matched band of sample and standard is removed from plates and dissolve in 2 ml methanol and keep overnight. Centrifuge for 10 minutes. The supernatent is collected twice by adding 2 ml and 1ml methanol. This final filtrate is used to note the O.D. on the Spectrophotometer at 360 nm alongwith standard using methanol as a blank.

Calculations

The amount of aflatoxin B1 (in ppb) is computed by using the following formula:

$$\text{Aflatoxin B1 content}(\mu g/kg) = \frac{S \times Y \times Z}{W \times Z}$$

Where,

S = III or O.D. of aflatoxin Bi standard equal to that of sample material being evaluated on the plate.

Y = Concentration of aflatoxin Bi standard in ug/rnl.

V = Solventrequired (Ill) to dilute final extract.

W = Weight of sample in grams (DM basis)

Z = Sample extract spotted (Ill) to give fluorescence intensity equal to standard B1 (S).

Precautions

1. *Filtration of initial chloroform or acetone*: Water extract of sample should be carried out as quickly as possible.
2. For column clean up the quality of silica gel indicated in text should only be used. The recoveries of aflatoxin will vary depending upon the type of silica gel.
3. It is also desirable to carry out the extraction procedure with sample containing added quantities of aflatoxin.
4. Do not allow the plates to be exposed to UV light for a long time as UV light causes decomposition of aflaotoxin.

5. Suitable dilution of standards and sample extracts should be properly selected for comparing the fluorescence spots.
6. The entire analysis should preferebly be carried out with in 24 h.

Observations

Sample No.	Volume of Sample	Volume of Standard & their conc.	Comparison of fluorescence under UV light	OD of comparable sample and standard	Aflatoxin B1(ppb)
1					
2					
3					
4					
5					
6					
7					
8					

Reference

AOAC 1990. Methods of analysis. Association of Official Analytical Chernists.f lfi'" Edn.) Washington, DC, 20044.

6.14. Estimation of Ochratoxin

Principle

Ochratoxin of feed samples is extracted into hexane and chloroform methanol solution. This extract is dissolved in NaHCOa solution. The extracted chloroform is passed through a column containing anhydrous NaCI and Na2S04, then dried. Again residues are redissolved in chloroform and aliquots are used for estimation of ochratoxin by thin layer chromatorgaphy.

Appratus

Silica gel, applicator, 5 glass plates (20cm x 20 em), hot air oven, geltank, U.V. lamp.

Reagents

Hexane, chloroform, methanol, sodium bicarbonate, anhydrous sodium chloride, anhydrous sodium sulphate, standard ochratoxin, tolueneethylacetate, formic acid, triethylamine, ammonia.

Procedure

1. *Sampling*: Sampling for ochratoxin (OT-A) analysis of feed sample is done in triplicate as per the procedure laid down for aflatoxin Br (1813426: 1992).
2. *Extraction of OT-A from feed samples*: 25 g sample is double extracted, first with 200 ml hexane and then with 100 ml. chloroform-methanol (1:1) solution (Chu and Butz, 1970). The hexane extract is further extracted by liquid partitioning with 150 ml 1% NaHC03. Chloroform methanol extract is dried and dissolved in same NaHC03 solution used to extract hexane solution. Impurities were cleared by liquid partitioning (twice) with chloroform; solution is acidified to pH 2.0 and extracted into the chloroform. The extracted chloroform is passed through a column (22 x 300 mm) containing anhydrous NaCl (5 g) followed by another column (22 x 300 mm) containing anhydrous Na2804 (5g) and then dried. Residues are redissolved in known quantities of chloroform and aliquots are used for estimation by TLC. Chloroform is used as the solvent for ochratoxinA (OT-A). Standards are prepared in different concentrations viz., 5, 10, 100 and 200 ug/rnl.
3. *Preparation of TLC plates*: Silica gel G-(Qualigens), 60 g, is stirred vigorously with 120 ml of distilled water in a .beaker for 10-15 minutes, until satisfactory slurry is obtained. The thickness of the silica gel slurry is adjusted with addition of a few ml of distilled water, if required. The contents are transferred into an applicator, which is adjusted to coat 0.25 mm thick layer-of slurry. Five glass plates (20cm x 20cm) are coated immediately by drawing the applicator over them at a uniform speed in order to get a uniform layer of the slurry. The coated plates are kept at room temperature, in a dust free atmosphere, till the plates got settled (10 minutes). Then the plates are heated in an oven at 100°C for one hour and stored in a plate chamber. Before use the plate are activated at 105°C for 2 hours for OT-A.
4. *OT-A Estimation*:

a) Known graded quantities (/11) of sample solutions and OT-A standard solutions are spotted 2 em from the bottom of the chromoplate leaving at least 2 em margin from either side. The spots are kept 1.5 cm apart and spot diameter being not more than 5 mm. The plates, are developed in tolueneethylacetate-90% formic acid (5:4: 1, v/v/v - for OT - A), to a depth of less than 1 em. The tank is saturated with the solvent before placing the plate in it. The solvent is allowed to run 14-15 ern above the origin; the chromoplates dried in horizontal position, and observed under long and short wavelengths ultraviolet light, simultaneously, in dark chamber. The OT-A in the sample is identified by greenish blue characteristic fluorescence and Rf value is 0.65, check with the standards run along with the samples.

b) Fluorescence intensities of the spots of sample extracts with that of the standards are compared under U.V. lamp and the concentration of the standard that matches the sample fluorescence intensity taken for further calculations. In a mismatch, either the sample extract is diluted or the concentration of spotted standards is decreased or increased.

c) Plates developed for OT-A estimation, are sprayed with a freshly prepared 25% solution of triethylamine in methanol, or are hold over a concentration of ammonia. With either procedure, the OT-A spots will give blue fluorescence in UV. light. This confirms the presence of OT-A in the sample Visual comparison of sample spot with standard spot under UV light is repeated and averaged with previous results.

Calculation

Concentration of OT-A is calculated from' the following formula:

$$\mu g\,OT - A / kg = \frac{S \times Y \times V}{X \times W}$$

Where,

S = ul, standard which matched the sample,

y = Concentratio of the standard in ug/rnl,

V = Volume ul, of final dilution of sample extract,

X = uL of sample extract spotted giving fluorescence intensity equal to IS',

W = 's' sample extracted.

References

Chu, F.S. and Butz, M.E. (1970). Spectrophotofluorodensitometric measurement of ochratoxin A in cereal products. J. Assoc. Off. Anal. Chern., 53: 1253-1257.

IS 13426 (1992). Animal feeds and feeding stuffs-methods of sampling for aflatoxin analysis, Bureau of Indian Standards, Manak Bhavan, Bahadur Shah Jafar Marg, New Delhi.

□□□

Chapter 7

Rumen Biochemistry and Enzymes Assay

In ruminants, nutritional studies are very complicated, unlike nonruminants where it is quite simple. In non-ruminants, analysis of feed taken and faeces voided by the animal gives overall picture of digestion and nutrient availability to the animal. In ruminants, since it has a complete microbial digestive system (rumen) within the animal digestive system, the analysis of the feed taken and faeces voided will not express the whole processes in the GI tract of the animal. To have complete picture, it is necessary to study at the site of microbial digestion (rumen). For such study, samples from the rumen have to be drawn and analyzed for metabolites, microbial enzymes etc.

7.1. Collection of Rumen Sample and its Preservation

Sampling from dead/slaughtered animals

The rumen content or rumen liquor can be sampled by this technique in rare wild animals and from those animals which are slaughtered after the experimentation. The whole rumen content can be collected, mixed manually and a representative sample is taken out. For rumen liquor sample the contents can be squeezed through double layer of muslin cloth.

Sampling from fistulated animals

This is the easiest method of obtaining rumen content samples, once the animal's body has been operated. The best way of sampling rumen contents is removing the whole contents, mixing it properly and returning the contents immediately back. Care should be taken that the rumen contents are not outside the animal's body for a longer period. The rumen liquor can also be sampled with the help of a probe. The probe consists of a perforated tube connected with a rubber tube. The perforated tube is covered with a nylon cloth, so that only the liquid portion of rumen contents can be sucked out through the rubber tube.

Sampling from intact animals

The rumen liquor sample is drawn with the help of a stomach tube. The thickness and stiffness of tube should be such that does not injure the epithelial layer of the mouth. The thickness of tube for cattle and-sheep should not be more than 3-4 cm and 1-2 cm, respectively. This method is most suitable for sampling rumen contents of clinically sick animals.

Sampling and preservation

For biochemical analysis collect the sample in the flask kept in ice bath, but for enzyme estimation rumen liquor collection should be made at 39°C.

1. Filter the rumen liquor through four layers of muslin cloth (strained rumen liquor).
2. For biochemical analysis 2 drops of 20% H_2SO_4 is added to 20 ml of rumen liquor immediately after collection.
3. Acidified samples should be stored at -200C till analyzed.
4. For enzymatic studies rumen liquor samples should be processed immediately after collection.

7.2. Recording of pH

pH of rumen sample should be recorded immediately after collection. Hand-held pH meter or filter paper strips may help for immediate recording. If the laboratory is not far away, the sample can be brought at the earliest and pH needs to be recorded.

Bench-Top digital pH meter

Hand-held digital pH meter

pH paper strips with close range around 7.0

7.3. Rumen Protozoa and its Quantification

In rumen microbial eco-system, protozoa have a very important role to play, especially when the animals are fed on good quality and starch based diet, as these protozoa engulf starch grain and protect it from immediate fermentation resulting in low pH which deleterious for rumen bacteria. In fibre degradation also protozoa contribute as they also produce fibre degrading enzymes. Therefore for evaluating a diet, study on shift in protozoal population is also important.

Fixing and staining of rumen protozoa

Reagents: Methyl green formal saline (NaCl 0.85g + Methyl green 0.06g + Formaldehyde solution 10ml + Distilled water 90ml)

- Take out rumen liquor from the fistulated animal with the help of a wooden probe fitted with a nylon cloth sieve and by sucking with a syringe.
- Collect the strained rumen liquor in already flushed tubes/flasks with oxygen free gas (N2 or carbon dioxide) and close the tubes/flasks tightly with butyl rubber stoppers.
- Transport the sample to the laboratory at a temperature varying between 35-40 °C.
- Immediately put a drop of sample with the pipette of wide orifice and cover it with the cover Slip and examine live protozoa under microscope.
- For counting of protozoa fix the strained rumen liquor sample immediately reaching in laboratory, by mixing one volume of methyl green formal saline solution with one volume of rumen liquor
- Leave the sample at room temperature for overnight.
- The specimens stained with methyl green can be stored for a very long time (more than a year).
- Methyl green stains the nucleus in green colour while the other parts of the cell are not stained.

- Wash the haemocytometer with water and alcohol and let it air dry.
- Put two drops of stained rumen liquor on the two cavities of haemocytometer.
- Cover with cover slip avoiding formation of air bubble under the cover slip.
- After putting the coverslip, the thickness of haemocytometer cavity is 0.1mm.
- Examine the slide under microscope and if the number of protozoa per microscopic field is more than 20-25 cells, then dilute the sample with 30% glycerol.
- Prepare fresh slide and count 30 microscopic fields for protozoa, equally distributed on the slide.
- Calculate the number of protozoa/ml of rumen liquor by the following formula:

$$\text{Number of protozoa / ml rumen liquor} = \frac{(N \times A \times D)}{(a \times V)} = 3919.27 \times N \times D$$

Where:

N = Average number of cells/microscopic field

A = Area on slide on which the sample is spread (area of the cavity of haemocytometer)

D = Dilution of rumen liquor

a = Area of microscopic field

v = Volume of rumen liquor in the counting chamber

NB. The factor 3919.27 is calculated in the present set of microscopical examination

7.4. Volatile Fatty Acids (VFA)

Principle

In acidic medium volatile fatty acids remain in the acid form and by heating get vaporized. The vapors of VF A are condensed and titrated against standard alkali in the presence of indicator (Barnett, and Reid, 1956).

Apparatus

Balance, Markham apparatus, Heating mantle, Burette, Conical flask, Pipette, Ice bath.

Reagents

1. Oxalate buffer

   a) Potassium oxalate 10% - Take 109 potassium oxalate, dissolve it and make the volume to 100 ml with distilled water.

   b) Oxalic acid (5%) - Take 5g oxalic acid, dissolve it and make volume to 100 ml with distilled water.

   c) Mix soln. A and B.

   Or

   d) Saturated magnesium sulphate in IN H2S04

2. 0.0lN NaOH

3. Phenolpthalein indicator: O.lg phenolpthalein, dissolve in 50 ml ethanol and add to it 50 ml distilled water.

Procedure

1. Connect the Markham apparatus with a round bottom flask (2 lit.) half filled with water, kept on heating mantle (Fig 15.1).
2. Allow the water to boil.
3. Pipette 2 ml strained rumen liquor and than 2 ml buffer in the apparatus. Wash the funnel with distilled water.
4. Allow the sample to boil.
5. Collect 100 ml distilled VFA in 250 ml flask kept in ice bath.
6. Add few drops of phenolpthalein and titrate the distillate with O.OlN NaOH till pink colour develops (end point).
7. Read the volume of O.OlN NaOH used for titration on the burette.

Calculation

Milli equivalent or mmol TVFA/100 ml rumen liquor

$$\frac{\text{Vol. of NaOH used} \times \text{Strength of NaOH}}{\text{Vol. of rumen liquor taken}} \times 100$$

$$\frac{\text{Vol. of NaOH used} \times 0.01}{2.0} \times 100$$

Precaution

1. Boiling of sample should be very gentle.
2. Distillate should be collected in precooled flask and should be kept in ice bath till titrated. Rise in temperature will result in loss of volatile fatty acids from the distillate.
3. During titration, stirr the contents of the flask continuously.
4. End point is the point where pink color persist atleast for 30 seconds.

7.5. Total Nitrogen

Principle

When sample containing nitrogen is digested with H2S04. the total nitrogen converts into ammonium sulphate $(NH4)_2SO_4$. In the presence of alkali, ammonia is released form the ammonium sulphate and released ammonia is distilled and trapped in a known volume of standard acid, which is than back titrated with the standard alkali. The whole process is completed in three steps. The reactions during estimation are as follows.

Digestion

$$\text{Organic nitrogen} + \text{Cone. } H_2SO_4 = (NH4)_2SO_4$$

Distillation

$$(NH4)_2SO_4 + 2NaOH = Na_2SO_4 + 2NH_3 + 2H_2O$$

Trapping

$$2NH_3 + H_2SO_4 = (NH4)_2SO_4$$

Reagents and chemicals

Reagents and chemicals and the procedure are the same as detailed in Kjeldahl N analysis under analysis of proximate constituents.

Procedure

1. Digestion

i) Take 5 ml strained rumen liquor in a kjeldahl flask
ii) Add 10 ml concentrated H_2SO_4.
iii) Add 2-3 g digestion mixture.
iv) Keep the flask on digestion bench and allow gentle boiling. Bumping should be avoided.

v) When the solution becomes clear blue, then remove the flask from the digestion bench and cool it.

vi) Add 5 to 10 ml distilled water to the kjeldahl flask. Transfer the whole material in 100 ml volumetric flask with repeated washings of distilled water. Make the volume to 100 ml.

2. Distillation

i) Set the kjeldahl distillation assembly

ii) Take 10 ml 0.01N H_2SO_4 (A) in a conical flask and add 2-3 drops of indicator.

iii) Keep the flask under the condenser in such a way that the tip of the condenser should be dipped in acid, to avoid ammonia loss during distillation.

iv) Take 10 ml aliquot of digested sample and transfer it into the kjeldahl assembly.

v) Add 15-20 ml 40% NaOH to make the aliquot contents alkaline and put the stopper immediately.

vi) Allow distillation for 15 min.

3. Titration

i) Remove the flask after washing tip of the condenser with distilled water.

ii) Titrate the contents of the flask with standard O.OH,T ~,[aOH till the pink color develops (end point).

iii) Record the volume of alkali (B) used for titration on the burette.

iv) Run a blank using all reagents but no sample and following the whole procedure to estimate the nitrogen contents of the reagents, if any.

Calculation

$$\text{Nitrogen/100ml rumen liquor} = \frac{V \times 0.00014 \times D}{v \times A} \times 100$$

Where,

V = A-B

D = Dilution (Volume made in volumetric flask)

v = Initial volume of rumen liquor taken for the digestion

A = Aliquot taken (10 ml)

Total nitrogen of sample = Total nitrogen of sample - Total nitrogen of blank.

Precaution

1. Bumping should not be allowed during digestion. Beads can be used to avoid bumping.
2. During distillation, first flask containing acid should be arranged under the condenser of the distillation assembly and thereafter alkali should be added in the sample. Because by adding alkali ammonia releases immediately.
3. Sample should be digested completely and clear blue color is the indication of complete digestion.

7.6. TCA Precipitable Nitrogen

TCA precipitable nitrogen is the nitrogen corning from the true protein because by adding trichloroacetic acid, only proteins are precipitated. ese precipitated proteins are analyzed further for nitrogen contents by microkjeldahl method as described above.

Procedure

1. Take 5.0 ml rumen liquor in a centrifuge tube.
2. Add 5.0 ml of 20% TCA
3. Leave the tubes overnight.
4. Centrifuge the tubes at 2,000 rpm for 10 min.
5. Transfer the whole precipitate with repeated washings of distilled water in kjeldahl flask.
6. Proceed for digestion, distillation and titration as described above.

Soluble nitrogen: TCA precipitable nitrogen subtracted from total nitrogen gives soluble nitrogen contents of rumen liquor.

7.7. Ammonia Nitrogen

Phenol-Hypochlorite Assay for Ammonia

Adapted from Broderick and Kang JDS 63:64 (1980)

Phenol reagent

Dissolve .15 g of sodium nitroferricyanide (sodium nitroprusside) in 1.5 L of distilled H_20. Then add 33 mL (90% wt/vol) phenol (measured by a graduated cylinder) and mix thoroughly. Make solution up to 3 L with distilled H_20, and store in brown glass bottle.

Hypochlorite reagent

Dissolve 15 g of sodium hydroxide in about 2 L of distilled H_20. Then dissolve 113.6 g of disodium phosphate (heptahydrate - Na2HP04'7H20) in this solution with mild heating and mixing. After cooling, add 150 mL of commercial bleach (5.25% sodium hypochlorite) and mix.

Then make solution up to 3 L with distilled H_20. Filter solution through Whatman #1 filter paper, and store in polyethylene bottle protected from light.

Ammonia standard solution

A stock solution of 100 mM ammonia was prepared by diluting .6607 g of ammonium sulfate (dry ammonium sulfate overnight in $100^{G}C$ oven before use) to 100 mL with .1 N HCI.

Working standards of **1,2,4,6,** and 8 mM are made by diluting aliquots of the stock solution.

7.8. Gas Chromatography

Chromatography can be defined as a physical method for separation of components in a mixture. There are several types of chromatography, including paper, thin layer, liquid and gas. We will confine our present discussion to gas chromatography, as it has wide application to the analysis of organic samples commonly used in nutrition studies.

The basis of gas chromatography is the separation column. The column has both a stationary phase, which may be either solid or liquid on a solid support, and a carrier gas mobile phase. Gas liquid chromatography (GLC), which is most widely used, employs a liquid stationary phase on solid support. When a volatile sample is injected onto the column, it encounters two opposing forces. One force, the carrier gas, pushes the sample through the column, whereas the second force, the interaction of components with the stationary phase, retards movement of the sample through the column. Individual components of the sample can be solubilized partially in the liquid stationary phase or absorbed to the solid support. Physio-chemical aspects of each component determine extent of interaction with the stationary phase, and because components have different properties, they are eluted from the column at different rates, resulting in separation of components.

The efficiency with which a column separates materials is obviously of concern and depends in part on the column temperature and gas flow rate. Efficiency can be determined from mathematical calculations and can then be used to determine optimum column conditions.

7.8.1. Basic components of a gas chromatograph

Gas supply and flow regulators provide a means of obtaining a uniform rate of gas flow (Zweig and Sherma, 1972). The carrier gas chosen will vary with the detector used, but should be inert and inexpensive. Nitrogen gas is a popular choice when a flame ionization detector is used (Fritz and Schenk, 1979).

The injection port is a heated chamber into which a liquid sample is injected. Injection is accomplished with a syringe through a self-sealing septum (Zweig and Sherma, 1972). Normally microliter quantities of the sample are injected, and the injection port temperature is maintained at a level that will immediately volatilize the sample. The flow of carrier gas then pushes the sample onto the column.

The column is held in an oven that maintains the desired temperature for analysis (Fritz and Schenk, 1979). Columns may be constructed of glass or metal (usually stainless steel). Column packing (solid support and liquid phase) varies depending on the nature of the substance analyzed. Several commercial firms are capable of assisting the analyst in choosing the proper column and typically provide pre-packed columns or column packing material for purchase. Length and diameter of the column also vary depending on the nature of the instrument and sample to be analyzed.

The detector sits at the end of the column. As components of the sample exit from the column, the detector functions to sense their presence and send out a signal proportional to the concentration of the component in the carrier gas (Fritz and Schenk, 1979). Different kinds of detectors are used depending on the nature of the material being analyzed. The detector also is heated to maintain sample components in a gaseous state.

The three most widely used detectors are the flame ionization detector, the thermal conductivity detector, and the electron capture detector. The *flame ionization detector*, is essentially a micro-burner surrounded by an electrode that collects ions produced when a sample component is burned (Fritz and Schenk, 1979). The flame is usually provided by a flow of hydrogen and air into the detector, and as an organic sample enters the flame with carrier gas, it is burned, resulting in the formation of ions. These ions migrate to the electrode, and the electrical current is amplified and sent to the recorder.

The thermal conductivity detector works on the principle that the rate of heat loss of a hot wire is proportional to the molecular weight of the gas surrounding the wire (Fritz and Schenk, 1972). If Helium (MW = 4) is used as a carrier gas, sample components in gaseous form will be of higher molecular weight and thus decrease thermal conductivity. Heated metal filaments in the detector form a Wheatstone bridge that measures the differences in thermal conductivity of pure carrier gas and carrier gas plus sample components. As a sample component passes over a filament, its temperature increases because of the lower thermal conductivity of the sample component compared with the carrier gas. This heating increases the resistance of the Wheatstone bridge, and the resistance needed to restore the balance between the two filaments (carrier gas filament and carrier gas + sample filament) is measured (Fritz and Schenk, 1979).

The electron capture detector is very useful for detecting organic halogen compounds and finds considerable use in pesticide analysis. The detector contains a p-emitting radioactive substance (usually 63_{N}i). Carrier gas reacts with p particles, producing a stream of electrons. Electrons are attracted to an anode in the detector, providing a baseline electric current. As sample components come off the column, they interact with p particles and decrease the baseline current, providing a signal that can be detected.

The amplifier and recorder are used to multiply detector response and provide a permanent record of detector output, respectively. Many new gas chromatographs have mini-computer attachments that intergrate peaks as they come off the column and actually calculate concentrations of components. Such computational instruments have largely eliminated the use of recorder-produced chromatograms in quantitative analysis of OLC data.

A comparison of detectors used in gas chromatographs

Detector	Output	Uses
Flame ionization	Output porportional to ions produced when sample is burned in H_2 + air flame.	Used for organic compounds. No response to CO_2, H_2, CO, H_2S, and NH_3. Useful for analysis of water extracts.
Thermal conductivity	Output is difference in thermal conductivity between pure carrier gas and carrier gas plus sample	Used for both inorganic and organic compounds. Sensitive to changes in gas flow and termperature.
Electron capture	Output proportional to change in current of carrier gas/p particle interaction caused by sample components.	Used for 0-, P-, S-, $N0_2$- and halogen-containing compounds. Weak response to ethers and hydrocarbons. Use ful in pesticide and insecticide analysis.

Other important terms in GLC

Retention time is the time lapse between sample injection and emergence of any component. Retention time varies depending on column packing, temperature, and gas flow rates. It is important that sample components have distinct retention times so separation can be effective. Generally, sample components are identified by comparing retention times of standard samples with retention times of unknown sample components.

Peak area is quantitatively related to the amount of component present in a sample and is threby of considerable importance. As indicated previously, peak area can be determined with the aid of computerized intergrators. It also can be determined manually by a process known as triangulation. This process entails transforming a peak into a triangle and determining peak area as peak height times peak width at half height.

Resolution is a term that indicates degree of separation. It depends primarily on the column packing but also may be influenced by gas flow and column temperature. Resolution is indicated by degree of peak separation and sharpness; good resolution is highly desirable in any GLC analysis.

7.8.2. Quantitation of GLC data

As suggested previously the nature of components in an unknown sample can be determined by comparing their retention times with the retention times of known components in a standard mixture. Once it is known "what the component is", peak area can be used to determine the concentration of components in the sample.

The basic idea behind quantitative GLC analysis is that peak area is proportional to the concentration of components in the sample. However, detector response varies with different components, and slight corrections are needed to ensure that a given peak area is equivalent to a given molar concentration. Usually this correction is small, and often ignored; however, computerized integration systems are usually capable of correcting for variations in detector response. A further problem in quantitation of GLC data is that variations in temperature, gas flow, and so on affect peak area. For this reason, most quantitative techniques employ some type of internal standard addition to adjust for such variations.

Thus, *GLC* is extremely important in quantitative analysis of a number of organic compounds. Applications include the measurement of fatty acids, amino acids steroids, pesticides, and a large number of other biologically important compounds.

Sample and standard preparation

1. Centrifuge strained ruminal fluid at 10,000 x g for 10 min.
2. Mix 5 mL of ruminal fluid supernatant with 1 mL of meta-phosphoric acid-2EB solution.
3. Stand in cold (ice bath) for> 30 min.
4. Centrifuge for 10 min. at 10,000 x g.
5. Take supernatant fluid for GLC injection.
6. *Standard preparation* - to ensure 2-EB amount is the same in standard and sample, take 5 mL of standard prepared in Section I and add 1 mL of the metaphosphoric:2-EB solution. Mix completely, and keep in cold.

7.8.3. Some common description for an overview of gas chromatograph

1. Why we have the technique

a. SEPARATE components and impurities
b. Allow QUANTITATION- need a standard
c. Allow IDENTIFICATION- need a standard

2. Requirements of the analyte for GC analysis

a. Volatile

b. Thermally stable

c. Non-reactive with the components of the instrument

3. Components of the technique

a. Injector- volatilize the analyte

b. Column- allow for a path from injector to detector

c. Detector- allow for detection and quantitation

d. Mobile Phase- normally Helium, though H_2, N_2, CO_2 etc. are acceptable

e. Stationary Phase- a coating on the capillary column, or packing on a packed column.

4. Means to affect the analysis (in order of importance)

a. Mobile phase (gas selection)

b. Column stationary phase selection (polarity, phase ratio, etc.)

c. Stationary phase loading (film thickness)

d. Oven temperature and/or temperature program

5. Important rules to remember

a. Precondition the column after lengthy down-time. Set oven ~50° C above normal oven temperature of your run; not to exceed MAX oven temperature of the column. Precondition overnight is preferred.

b. Keep injector temperature above analyte boiling point.

c. Keep injector temperature ~50° C above oven temperature.

d. Keep detector temperature ~50° C above oven temperature.

e. Don't inject aqueous samples.

f. Air in the column at elevated temperatures (>100° C) will result in column damage.

6. Calculating linear velocity (ideally around 30 cm/sec) and column flow through the column

a. Inject something not retained by the column- i.e. acetone vapor with oven @ 100° C.

b. Note time of first peak.

c. Linear velocity = Column Length (cm) / retention time of unretained component

d. Volume of a cylinder = p r^2 h

e. Column flow = p r^2 h / retention time of unretained component

7. Notes for sample analysis

a. Sample Concentration should be ~1 m g/mL (1 ppm) since you will be using SPLITLESS INJECTION- meaning that the entire sample injected goes onto the column.

b. Instrument Preparation

i. Verify instrument prepared by the TAs by noting:

1. Instrument READY light illuminated. If NOT READY, press CLEAR to display reason for NOT READY.
2. Clicking noise of the TCD detector.
3. Column Head Pressure gauge somewhere between 5-10 psi.

ii. Verify your instrument parameters.

1. Oven Temperature
2. Injector A Temperature
3. Detector A Temperature
4. TIME for next run (press TIME until "NEXT RUN: xx minutes" appears)

iii. Verify integrator ready

1. Press LIST TIME for stored time events (Run Time and Integrator Inhibits).
2. Press LIST twice for a printout of standard parameters.

c. Injection technique suggestions

i. Purge syringe with 5 syringe barrel volumes to remove previous sample/rinsate.

ii. Slowly draw 2 m L of sample, then adjust volume to 1 m L.

iii. Wipe syringe needle with Kimwipe.

iv. Draw in 1 m L of air.

v. When READY light is illuminated on the GC, inject the sample into the Injector

1. Support plunger by gentle pressure on the side (keep plunger from ejecting due to column head pressure.
2. Support needle on penetrating the septa to keep it from bending.
3. On full insertion of the syringe, immediately depress the plunger (without bending it).
4. Push START on the GC control panel.
5. Withdraw the syringe.
6. Minimize time for penetrating septa to withdraw of the syringe.

vi. For best results (i.e. least amount of cross-sample contamination), rinse syringe 5 times with applicable solvent. This step is normally not required for CHEM 337.

d. Post-analysis suggestions

i. Peaks integrated satisfactory with baseline resolution.

ii. Peak area counts not off scale; this would be denoted with a symbol on the integration results table.

iii. All peak widths representative- really wide peaks may be from someone else's previous injection.

iv. Record instrument parameters, particularly column type and dimensions; column head pressure; temperatures (injector, oven and detector); and detector type (TCD).

v. Integrator Attenuation will not affect area counts or results- just the plot.

7.9. Rumen Microbial Enzymes

7.9.1 Handling of enzymes

Enzymes are very fragile substances and very soon undergo denaturation and inactivation under unsuitable conditions. Precautions should be taken to protect the enzymes from inactivation during handling them. Although the suitable conditions vary with different enzymes but there are a few points which are applicable to most of the enzymes.

Temperature

High temperatures are always detrimental for the enzymes. Care should be taken not to expose the enzymes above body temperature, even at body temperature not for a longer time. During the process of isolation and purification a temperature of about 4°C should be maintained.

pH

Highly acidic and alkaline conditions are destructive for the enzymes. In general, majority of enzymes are inactivated below pH 5.0 and above pH 9.0, though exceptions are there. Thus while adjusting pH by adding acid or alkali, the drop should go slowly down the side of the vessel, with gentle and continuous stirring of the solution to avoid formation of destruction zones around the drops. It is always preferred to maintain pH at neutral during handling the enzymes.

Froth formation

Enzyme preparation should be transferred very carefully to avoid frothing because many enzymes denature at the surface. Thus the enzyme solution should be poured by the side of the vessel to avoid entry of air bubbles.

Organic solvents

Organic solvents like alcohol, acetone etc. inactivate the enzymes. Use of organic solvents should be avoided but if requried, precaution should be taken while adding it and should be added at low temperature (-15 to -20°C) and very slowly with continuous and gentle stirring.

Detergents

Use of detergents should be avoided as they also inactivate the enzymes. Especially anionic detergents like sodium docecyl sulphate (SDS), sodium deoxycholate (DOC) or cationic detergents like cetyltrimethylammonium bromide etc. should not be used. Some times a nonionic detergent like triton x 100 is used for breaking the cells to release the enzymes. It should be used at a very low concentrartion with short treatment time.

Storage

The enzymes should be stored in deep-freeze where they are stable even for a few months. Repeated freezing-thawing should be avoided because it results in loss of activity. Sometimes enzymes can be stored in a precipitated form in saturated ammonium sulfate solution. When required, they can be centrifuged down and dissolved in water. It is always recommanded to store the enzyme preparation in small aliquots for one time use. Enzymes in concentrated form are more stable than in diluted form.

Bacterial contamination

Enzyme preparations should be protected from bacterial contamination while storing for a long period by adding protectants like sodium azide etc.

7.9.2. Practical precautions during enzyme estimation

While estimating the enzyme activity, to maintain linearity of the reaction, the conditions in the reaction mixture must be maintained as constant as possible.

Temperature

Reaction should be carried out in thermostatic conditions. All the reagents to be used for the reaction must be brought at a required temperature before mixing them to start the reaction. For each enzyme the optimum temperature for its maximum activity is different, so the enzyme reaction should be carried out at optimum temperature for the enzyme, to get maximum activity. If not possible then carryout the enzyme reaction at ambient temperature.

Buffering system

Activity of an enzyme is maximum at a particular pH called optirnium pH for that particular enzyme and little change in the pH, changes its rate of reaction. Therefore when an enzyme activity is estimated, a recommended p H should be maintained throughout the reaction, and for that proper buffereing system has to be provided for the ass say mixture to prevent the change in pH during the reaction.

Contamination

Enzymes are highly catalytic in nature so any contamination in the vessel should be avoided which can stimulate or inhibit the reaction. ego in case of amylase estimation, slight contamination of saliva during pipetting in the reaction mixture will increase the enzyme activity many folds. Thus pipetting should be done very carefully and glassware should be washed properly.

Reaction rate

Reaction rate should not be very fast or slow and should be adjusted in such a way that the readings could be made properly and with in the range. The reaction rate can be adjusted by increasing or decreasing the volume of the enzyme in the reaction mixture. Care should be taken that enzyme volume should not be very large in assay mixture. It is preferable to have enzyme preparetion in the concentrated form so that very small quantity of the enzyme sample may be sufficient to get reaction within the range.

Mixing

When the reaction is started either by adding substrate or enzyme in the assay mixture, immediate thorough mixing is required because reaction starts immediately as soon as substrate and enzyme come in contact with each other.

7.9.3. Fraction of rumen liquor for enzyme estimation

Reagents

1. Phosphate buffer 0.1 M, pH 6.8
   Solution 1: Dissolve 17.799 g Na2HP04. 2H20 to make the volume 1 litre. Solution 2: Dissolve 15.601 NaH2P04. 2H20 to make the volume 1 litre. Take solution 1. And adjust pH to 6.8 by adding solution 2.

Procedure

Rumen liquor is collected through a permanent fistula in the rumen or with the help of a stomach tube. The frequency of rumen liquor collection may vary with the type of study and in most cases, it is at 4 h post feeding for enzyme estimation. The collection should be done at 39^0e ± 2°e in a conical flask and plug it tightly. Transport the flask immediately to the laboratory for processing.

Protozoa rich fraction

Take 20 ml of thoroughly mixed SRL and centrifuge at 450 x g for 5 min at 37°C. Decant the supernatant and store it in a labeled test tube for further fractionation. The pellet formed in the centrifuge tube is suspended in the starting volume (20 ml) of 0.1 M phosphate buffer pH 7.0. Centrifuge again at 450 x g for 5 min at 4°C and remove the supernatant. Repeat the process twice and suspend the finally washed pellet in 10 ml of phosphate buffer.

Bacteria rich fraction

The supernatant collected after the removal of protozoa is centrifuged at 27000 x g at 370C for 30 min. The entire supernatant is removed and used as cell-free rumen fluid. The bacterial pellet left in the centrifuge tube is washed thrice with phosphate buffer as in the case of protozoa but centrifugation is done each time at 27000 x g at 4°C for 30 min. The washed bacterial pellet is suspended in 10 ml phosphate buffer.

Disruption of protozoa and bacteria rich fractions

The protozoa and bacteria rich fractions kept in ice bath, are disrupted by sonication at 1011 for 5 min with 30 sec break. It is followed by centrifugation at 27000 x g for 30 min at 4°C. The supernatant of each fraction is used as the source of protozoal and bacterial enzymes.

Measure the volume of cell free fraction, protozoa rich fraction and bacteria rich fraction (A, B and C) and estimate the protein concentration.

7.9.4. Determination of rumen microbial enzymes

The extraction of enzymes from microbial biomass and the estimation of enzymes following the method as described by Agrawal et al. (2000) with slight modifications. Ten millilitre of strained rumen liquor (SRL) was centrifuged at 14000 rpm for 20 min., supernatant and pellet were collected separately. The supernatant was used as enzymes source for extra cellular portion. The pellet containing microbial biomass (bacteria, fungi and protozoa) was suspended in a solution containing 5 ml 0.1 M phosphate buffer (pH 6.8), 2 ml of carbon tetra chloride and 2 ml of 0.4% lysozyme. The suspension was incubated for 3 h at 39°C and then centrifuged at 14000 rpm for 20 min. The supernatant was collected and used as enzymes source for cellular portion. The activities of the enzymes viz. carboxy methyl cellulase (CM), α- amylase, microcrystalline cellulase (MCCase), filter paper (FP) degrading enzyme, α-glucosidase were determined separately for individual animal.

For the estimation of CMCase and α - amylase, the reaction mixture contained 0.1M phosphate buffer (pH 6.8) 1 ml, carboxy methyl cellulose (1%) 0.5 ml; buffer 0.5 ml, soluble starch (1%) 0.25 ml, enzyme 0.25 ml and incubated for 60 and 30 min respectively. For the estimation of MCCase, the 1% substrate (microcrystalline cellulose) suspension in the buffer was incubated at 10°C for 48 h prior to the use in the enzyme assay. The 2 ml assay mixture contained 1ml substrate suspension and 1ml enzyme solution and incubated for 60 min, with intermittent shaking. For the estimation of FP degrading enzyme activities, the assay mixture contained 1ml buffer, 50 mg whatman No.1 filter paper and 1ml enzyme with the incubation of 60 min with intermittent shaking. In the all above assays, the reaction was stopped by adding 1 ml of dinitrosalicylic acid solution. The reducing sugars liberated due to the enzyme activities were estimated by dinitrosalicylic acid reagent (Miller, 1959). The enzyme activities were expressed as μmole of reducing sugars produced per minute per ml under the assay conditions.

Reagent

1. Phosphate buffer – 0.1M pH-6.8
2. Lysozyme solution – 0.4 % in phosphate buffer
3. Carbon tetra chloride
4. Carboxy Methyl Cellulose: Dissolve 1g CMC in 50 ml distilled water by heating and make the volume up to 100 ml.
5. Microcrystalline cellulose: Dissolve 1 g MCC in 50 ml phosphate buffer and make the volume to 100 ml, incubate at 10°C for 48 h before use.
6. Di-Nitro Salicylic Acid Solution (DNS): Dissolve 10 g of NaOH pellets in 500ml distilled water. Add 10 g of DNS and add 2 g phenol, dilute the solution up to 1000 ml with distilled water.
7. Rochelle salt (40%): Dissolved 40 g Potassium Sodium Tartrate in distilled water and make the volume up to 100 ml.
8. Standard glucose solution (0.1%): Dissolved 0.1g glucose in distilled water and make the volume up to 100 ml.
9. Avicel (1%): Dissolved 1g avicel in 100 ml phosphate buffer.
10. Starch (1%): Dissolved 1g starch in 50ml distilled water by heating and make the volume up to 100 ml.
11. Xylan (1%): Dissolved 1g xylan in 100 ml distilled water and heat it at 70° C for 14 min. Centrifuge it at 3000g for 10 min and use the supernatants the substrate for xylanase.

12. Xylose standard (0.1%): Dissolved 100 mg in 100 ml distilled water.
13. Casein (1%): 1g of Casein fat free substrate was dissolved minimum volume of 1N NaOH, pH was adjusted to 7.5 with 1N HCl (drop by drop with mixing) and volume was made to 100 ml with phosphate buffer.
14. p-Nitrophenyl acetate: 0.0362 g of substrate was dissolved in 100 ml DMSO
15. Dimethyl sulphoxide (DMSO)
16. BSA standard: 0.06 g BSA was dissolved in 100 ml distilled water.
17. TCA (20%): Dissolved 20g freshly weighted TCA and in 100ml of distilled water
18. Solution A: Dissolve 2 g Na_2CO_3 in 100 ml of 0.1N NaOH solution and stored at room temperature.
19. Solution B: Dissolve 1g sodium potassium tartarate in 100 ml distilled water. To it 0.5 g copper sulphate was added and keep for overnight and then filtered to remove the precipitate, if any. The solution was stored at room temperature for later use.
20. Solution C: Mix 50ml solution A and 1ml solution B just before use.
21. Solution D: Mix 1 ml of Folins & Ciocalteu's phenol reagent and 2ml of distilled water.

Protein estimation in SRL (EC & C fractions)

Reagents

i) BSA solution (0.06 %): 60 mg/100 ml in distilled water
ii) TCA 20% solution: Dissolve 20 g TCA in 100 ml distilled water
iii) Solution A: Dissolve 2 g sodium carbonate in 100 ml 0.1 N NaOH
iv) Solution B: Dissolve 1 g sodium potassium tartrate in 100 ml distilled water. Add 0.5 g copper sulphate keep overnight and then filter to remove precipitate
v) Solution C: Mix solution A 50 ml with 1 ml solution B just before use
vi) Solution D: Mix 1 ml Folin and Ciocalteu's phenol and 2 ml distilled water just before use

*Solution A & B can be stored at RT

10 ml SRL

Centrifuge 14000 rpm, 20 min

Collect supernatant
(Extra cellular portion for enzyme estimation)

Pellet

To the pellet add followings

- 5 ml 0.1 M Phosphate buffer (pH 6.8)
- 2 ml carbon tetra chloride
- 2 ml 0.4 % lysozyme

Incubate for 3 h at 39°C

Centrifuge 14000 rpm, 20 min

Collect supernatant
(Cellular portion for enzyme estimation)

Fractionation of Cellular and extra cellular portions of SRL for enzyme estimations

Estimation of protein

	Standard											Test
	1	2	3	4	5	6	7	8	9	10	11	EC/C
BSA con. in std (μg)	00	30	60	90	120	150	180	210	240	270	300	
Std BSA solution, ml	0.0	0.05	0.10	0.15	0.20	0.25	0.30	0.35	0.40	0.45	0.50	-
DW, ml	0.50	0.45	0.40	0.35	0.30	0.25	0.20	0.15	0.10	0.05	0.00	0.40
Sample (EC/C), ml	-	-	-	-	-	-	-	-	-	-	-	0.10
Soln C, ml	5.0	5.0	5.0	5.0	5.0	5.0	5.0	5.0	5.0	5.0	5.0	5.0
Incubate the tubes for 10 min at RT												
Soln D, ml	0.5	0.5	0.5	0.5	0.5	0.5	0.5	0.5	0.5	0.5	0.5	0.5
Mix well immediately and record the absorbance at 600 nm against blank												

Protein precipitation

1. Take 1 ml sample and add to it 1 ml 20 % TCA solution, incubate tubes 1h at RT
2. Centrifuge the tubes at x450 g for 10 min.
3. Collect the precipitate and dissolve in 1 ml 1 N NaOH solution and use for protein estimation

Estimation of Protease activity

		Test	Control	Blank
1	Phosphate buffer 0.1M, 6.8 pH	1.5 ml	1.5 ml	1.5 ml
3	SRL (EC or C)	0.25 ml	0.25 ml	-
4	Distilled water	-	-	0.25 ml
5	TCA (20 %)	-	2.0 ml	2.0 ml
	Substrate (Casein)	0.25 ml	0.25 ml	0.25 ml
	Mix well and incubate at 39 °C for 120 min			
6	TCA (20 %)	2.0 ml	-	-
	Keep overnight at room temperature			
	Centrifuge at 5000 rpm, 10 min			
	Collect supernatant and estimate protein			

Enzyme activity (Units) = μg protein hydrolysed/min/ml

$$\frac{\mu g \text{ hydrolysed protein}}{T \times S}$$

Casein (1%) substrate : 1g fat free Casein dissolved in minimum volume of 1N NaOH, pH was adjusted to 7.5 with 1N HCl (drop by drop with mixing) and volume was made to 100 ml with phosphate buffer.

Protocol for carboxymethyl (CMCase) cellulase

		Test	Control	Blank
1	Distilled water	-	-	2.0 ml
2	Phosphate buffer 0.1M, 6.8 pH	1.0 ml	1.0 ml	-
3	C M Cellulose	0.5 ml	0.5 ml	-
4	SRL (EC or C)	0.5 ml	0.5 ml	-
5	DNS*	-	3.0 ml	3.0 ml
	Mix well and incubate at 39 °C for 60 min			
6	DNS*	3.0 ml	-	-
	Keep in boiling water bath for 10 min			
7	Rochelle salt	1.0 ml	1.0 ml	1.0 ml
8	Distilled water	14.5 ml	14.5 ml	14.5 ml
	Mix well			
	Read the absorbance at 575 nm against blank			

Prepare a standard curve of Glucose (CMCase & MCCase & amylase)

Tube No.	1	2	3	4	5	6	7	8	9
DW (ml)	2.00	1.75	1.50	1.25	1.00	0.75	0.50	0.25	0.00
Glucose std (ml)	0.00	0.25	0.50	0.75	1.00	1.25	1.50	1.75	2.00
DNS (ml)	3.00	3.00	3.00	3.00	3.00	3.00	3.00	3.00	3.00
Mix well and keep and boiling water bath for 10 min									
Rochelle Salt (ml)	1.00	1.00	1.00	1.00	1.00	1.00	1.00	1.00	1.00
DW (ml)	14.5	14.5	14.5	14.5	14.5	14.5	14.5	14.5	14.5
Measure the OD at 575 nm, plot the curve of OD against glucose concentration									

N.B. For MCCase: The 2 ml assay mixture contained *1ml substrate* suspension and *1ml enzyme solution* and incubated for *60 min*, with *intermittent shaking*.
*Dissolve 0.05% Sodium sulphite in DNS just before use.

Estimation of α- amylase

		Test	Control	Blank
1	Distilled water	-	-	1.0 ml
2	Phosphate buffer 0.1M, 6.8 pH	0.5 ml	0.5 ml	-
3	Starch	0.25 ml	0.25 ml	-
4	SRL (EC or C)	0.25 ml	0.25 ml	-
5	DNS*	-	3.0 ml	3.0 ml
	Mix well and incubate at 39 °C for 30 min			
6	DNS*	3.0 ml	-	-
	Keep in boiling water bath for 10 min			
7	Rochelle salt	1.0 ml	1.0 ml 1.0 ml	
8	Distilled water	14.5 ml	14.5 ml	14.5 ml
	Mix well			
	Read the absorbance at 575 nm against blank			

*Dissolve 0.05% Sodium sulphite in DNS just before use.

Protocol for estimation of Xylanase

		Test	Control	Blank
1	Distilled water	-	-	2.0 ml
2	Phosphate buffer 0.1M, 6.8 pH	1.0 ml	1.0 ml	-
3	Xylan	0.5 ml	0.5 ml	-
4	SRL (EC or C)	0.5 ml	0.5 ml	-
5	DNS*	-	3.0 ml	3.0 ml
	Mix well and incubate at 39 °C for 30 min			
6	DNS*	3.0 ml	-	-
	Keep in boiling water bath for 10 min			
7	Rochelle salt	1.0 ml	1.0 ml	
8	Distilled water	14.5 ml	14.5 ml	14.5 ml
	Mix well			
	Read the absorbance at 575 nm against blank			

Prepare a standard curve of Xylanase

Tube No.	1	2	3	4	5	6
DW (ml)	2.00	1.70	1.40	1.10	0.80	0.50
Xylose (ml)	0.00	0.30	0.60	0.90	1.20	1.50
DNS (ml)*	3.00	3.00	3.00	3.00	3.00	3.00
Mix well and keep and boiling water bath for 10 min						
Rochelle Salt (ml)	1.00	1.00	1.00	1.00	1.00	1.00
DW (ml)	14.5	14.5	14.5	14.5	14.5	14.5
Measure the OD at 575 nm, plot the curve of OD against Xylan concentration						

*Dissolve 0.05% Sodium sulphite in DNS just before use.

Estimation of β Glucosidase activity

		Test	Control	Blank
1	Distilled water			2.0 ml
2	SRL (EC or C)	0.2 ml	0.2 ml	-
3	Sodium carbonate	-	2.0 ml	2.0ml
4	PNPG	1.8	1.8 ml	-
	Mix well and incubate at 39 °C for 10 min			
5	Sodium carbonate	2.0 ml		
	Mix well			
	Read the absorbance at 400 nm against blank			

Prepare a standard curve of p-nitrophenol

Tube No.	1	2	3	4	5	6	7	8
Con. p-nitrophenol (ug/ml)	**0.0**	**2.0**	**4.0**	**6.0**	**8.0**	**10.0**	**15.0**	**20.0**
DW (ml)	2.00	1.96	1.92	1.88	1.84	1.80	1.70	1.60
p-nitrophenol (ml)	0.00	0.04	0.08	0.12	0.16	0.20	0.30	0.40
Sodium carbonate (ml)	2.0	2.0	2.0	2.0	2.0	2.0	2.0	2.0
Read the absorbance at 400 nm against blank								
Prepare calibration curve by OD against p-nitrophenol concentration								

Enzyme activity (Units) = μmol p-nitrophenol/ min/ml

$$\frac{\mu g\ p-\text{nitrophenol}}{T \times S \times 139.11} \text{ or } \frac{\mu g\ p-\text{nitrophenol}}{139.11}$$

Estimation of β Xylosidase activity

		Test	Control	Blank
1	Distilled water			2.0 ml
2	SRL (EC or C)	0.2 ml	0.2 ml	-
3	Sodium carbonate	-	2.0 ml	2.0ml
4	PNPX	1.8	1.8 ml	-
	Mix well and incubate at 39 °C for 10 min			
5	Sodium carbonate	2.0 ml		
	Mix well			
	Read the absorbance at 400 nm against blank			

Prepare a standard curve of p-nitrophenol

Tube No.	**1**	**2**	**3**	**4**	**5**	**6**	**7**	**8**
Con. p-nitrophenol (ug/ml)	**0.0**	**2.0**	**4.0**	**6.0**	**8.0**	**10.0**	**15.0**	**20.0**
DW (ml)	2.00	1.96	1.92	1.88	1.84	1.80	1.70	1.60
p-nitrophenol (ml)	0.00	0.04	0.08	0.12	0.16	0.20	0.30	0.40
Sodium carbonate (ml)	2.0	2.0	2.0	2.0	2.0	2.0	2.0	2.0
Read the absorbance at 400 nm against blank								
Prepare calibration curve by OD against p-nitrophenol concentration								

Enzyme activity (Units) = μmol p-nitrophenol/ min/ml

$$\frac{\mu g\ p-\text{nitrophenol}}{T \times S \times 139.11} \text{ or } \frac{\mu g\ p-\text{nitrophenol}}{139.11}$$

7.10. Rumen Fluid Volume & Its Flow Rate

Principle

Polyethylene glycol (PEG) is a high molecular weight compound which 18 inert to rumen fermentation. The concentration in the rumen liquor declines with the passage of digesta from rumen to lower G.!. tract and is proportional to time. Protein free filtrate of rumen liquor gives turbidity in the presence of trichloro acetic acid and barium chloride which is a measure of PEG concentration at different post feeding hours.

Reagents

1. Polyethylene glycol (PEG) solution/(lO%, w/v)

   25g PEG (PEG 6000) is dissolved in distilled water to make volume 250 ml

2. Barium chloride (2.5%, w/v): Weigh 2.5g Ba Clz and dissolve in distilled water to make volume 100 ml.

3. 0.3N Ba(OH)2

4. 5% 2nS04. 7H20. Dissolve 5g 2nS04. 7H20 in distilled water to make volume 100 ml

5. Trichloro acetic acid- Barium chloride solution: 5 g Barium chloride is dissolved in distilled water to make volume 100 ml. Then 25g trichloro acetic acid is added.

6. Standard PEG solution: Stock (1 mg/ml): weigh 100mg PEG and dissolve in distilled water to make final volume 100 ml.

7. Working standard (100 ug/ml.): Take 10 ml stock PEG solution and dilute with distilled water to make volume 100 ml.

Procedure

1. PEG solution (10%, w/v) is infused into the rumen at 4 different sites with the help of hard polythene tubes equipped with funnel on the top and multiple holes at the closed bottom, for dispersal of solution in all directions inside the rumen, The container and tube is given two washings for the transfer of PEG quantitatively.

2. The rumen contents are mixed thoroughly and strained rumen liquor (SRL) is collected from different sites at hourly interval (1,2,3,4,5,6,7,8h) upto 8h post feeding.

3. SRL is centrifuged at 5000 rpm for 15 min.

4. 2 ml supernatant SRL is taken in a test tube, into which 2 ml reagent 3, 2 ml reagent 4 and 4 ml reagent 2 are added.

5. The tube with the whole content (lOml) is shaken vigorously and allowed to stand for 5 min

6. It is filtered through whatman No. 42 filter paper.

7. The filtrate, blank and the standard tubes are prepared as per the protocol given below.

8. The tube with the contents are mixed thoroughly and the turbidity developed is measured at 540nm immediately after 20 min.
9. Plot a standard curve and the concentration of PEG at different hours post feeding (y) is plotted against time (x) to draw a regression line (y = a + bx). Concentration of PEG at Oh is obtained by extrapolating the graph to Oh.

Reagent	Blank	Sample	Standard					
Distilled Water (ml)	2.5	1.5	2.4	2.3	2.1	1.9	1.7	1.5
Standard PEG solution	-	-	0.1	0.2	0.4	0.6	0.8	1.0
Concentration of standard (μg)	-	-	10	20	40	60	80	100
Sample filtrate (ml)	-	1.0	-	-	-	-	-	-
Reagent (ml)	2.5	2.5	2.5	2.5	2.5	2.5	2.5	2.5

Calculation

$$\text{Rumen fluid volume}(1) = \frac{\text{Amount PEG inf used int o the rumen}(g)}{\text{Concentration of PEF at OH}(g / ml)} \times 1000$$

Rumen flow rate (t/h) = Rumen fluid volume (l) × exponential decline in PEG concentration (slope b)

References

Smith, R.H. (1970). Nucleic acid metabolism in the ruminant. *British J. Nutr.* 24: 545-556.

□□□

Chapter 8

Blood Biochemical Assay

8.1. Blood Metabolites

In domestic farm animals nutritional imbalances and excesses is very much correlated with the turnover of fluids, salts and soluble organic materials in the system which is reflected upon the secretions and excretions from the body and the internal circulatory medium. The volume of changes in intake and secretion and the rapidity with which they can occur that reduce the metabolic stability of animal. The production diseases which usually related to active growth, early lactation, parturition, high milk production etc., rapid rate of exchange of water, sodium, calcium, magnesium, chlorides, phosphates and other organic nutrients is associated with rapidity in metabolic turnover and incompetence there in. When the imbalance is maintained, it may lead to change in amount of the body's reserves of certain metabolites, or their thoroughput leading to production diseases and are well reflected in blood.

Blood plays many important roles in coordinating the individual cells into a whole complex organism, which is accomplished by the presence in the fluid of dispersed or dissolved nutrients, metabolites, electrolytes, hormones, substances to counteract infection and haemorrhage and by equilibra between the cell and the blood stream so that homeostasis with respect to temperature, oxidation-reduction potential and ionic concentration is maintained throughout the organism. Thus it serves as the principal transport medium of the body, carrying 02, nutrients and chemical messengers (enzymes, hormones) to the tissues and waste products and synthesized metabolites away. General composition of blood is summarized below:

Blood

Vol. 5.7% ofBW, Sp. Gr. 1.060, pH 7.38, Viscosity 1.7-2.0

Freezing pt. 0.56°C, Os. Pr. 7.6 Hg at 37°C

Cellular fraction Vol. 45%	**Plasma fraction** Vol. 55%, S p.Gr. 1.026	
Constituents cells	Diffusible constituents	Non-diffusible constituents
1. Erythrocytes 2. Leukocytes 3. Platelets	1. Catabolic products: urea, creatinine, uric acid etc. 2. Anabolic c onstituents: glucose, amino acids, creatine etc. 3. Electrolytes: Na^+, K^+, Ca^{++}, Mg^{++}, Cl^-, HCO_3^-, HPO_4^-, $H_2PO_4^-$, etc. 4. Hormones, Vitamins etc	(Proteins and Polypeptides) 1. Albumins 2. Globulins 3.Fibrinogen 4. Enzymes, Lipids, etc

The functions of various constituents of blood and the -diagnostic role of such fractions relating to nutritional deficiencies or imbalances are given in the following table:

8.2. Blood Profiles as Indicators of Nutritional Status

A "metabolic profile" is defined as a series of specific analytical tests run in combination and used as a diagnostic aid. Blood tests from individual animals are routinely used to diagnose disease problems in dairy cattle. Veterinarians, producers and nutrition consultants alike seem to be interested in extracting pertinent information relative to herd nutrition and health status from blood tests. The Compton Metabolic Profile (CMP) has traditionally been used in this approach to:

1. monitor metabolic health of the herd;
2. help diagnose metabolic problems and production diseases
3. identify metabolically superior cows.

The CMP involved collecting 7 - 10 blood samples from 3 predefined groups of dairy animals, i.e., dry, peak lactation and midlactation, and having selected metabolites measured. From the test results, averages for each metabolite were calculated for each respective group and compared to reference values. Seven animals are considered the minimum number sampled to be statistically significant for interpretation. The goal of any metabolite profiling is to obtain the "population" mean and determine dynamic changes over transitions. Blood analyses are often used in lieu of other more appropriate diagnostic procedures such as ration evaluation and physical exams and without regard for proper technique to ensure sound diagnostic information. However, it can reveal some useful information if properly interpreted in conjunction with animal and ration evaluations. The combination used within a team approach can be an extremely useful diagnostic tool in nutritional evaluations of the dairy herd. It is only when the whole picture is evaluated will the uses of metabolic profiles produce useful diagnostic information. Thus, some of the important criteria need to be looked in to are:

- Metabolic profiles can be used to assess individual animal or herd nutritional status when used in conjunction with animal and ration evaluations.
- To assess a fresh cow problem, one should collect samples from far off dry, close-up dry, and fresh cows.
- Pooled blood samples can be used to minimize costs.
- Interpretation of pooled samples cannot be accomplished in the same method as with individual samples.

Interpretation of energy, protein, micronutrient status in conjuction with liver and kidney function test provides a whole lot of useful information in diagnosing a herd metabolic profile and are discussed below.

8.2.1. Assessment of energy balance

Energy balance is by and far one of the most critical nutritional factors impacting on animal health, lactation, and reproductive performance. Traditionally we have monitored changes in energy balance via body weight and condition score changes over time. This procedure may not be a sensitive enough tool when dealing with the transition cow. However, body condition score monitoring is still an important management tool, especially in assessing body condition changes with lactational performance. Another parameter that might be useful in assessing energy status is ketone body concentrations. At present measurement of beta-hydroxybutyrate (BOHB) concentration is most commonly used. However BOHB concentrations may not be sensitive enough and can come from dietary sources. A third method is a traditional research procedure, which has recently received much interest in the field. This is measurement of nonesterified fatty acids (NEFA) as a determination of energy balance. Many research studies have shown good correlations between energy balance and serum NEFA concentrations. Serum NEFA concentration is the result of adipose tissue breakdown of fat in response to negative energy balance. Circulating NEFAs are absorbed and metabolized for energy by the liver and other tissues. Concentration of NEFA then directly reflects the amount of adipose (fat) tissue breakdown taking place. Excessively high NEFA concentrations due to negative energy balance results in fatty infiltration of the liver, which is associated with higher incidence of periparturient metabolic diseases. Clinical experience suggests serum NEFA concentrations to be more sensitive to energy balance changes compared to body condition scoring in transition cow situations.

8.2.2. Assessment of protein nutrition

Assessing protein status is a bit more difficult than energy balance. At present there is no single metabolite that can be measured, which directly reflects protein status. As a result, a combination of parameters needs to be utilized, including blood urea nitrogen (BUN), creatinine, total protein, albumin, and creatine kinase (Ck). Urea nitrogen concentrations are influenced by a wide variety of interrelated parameters including: dietary protein intake and rumen degradability; dietary amino acid composition; protein intake relative to requirement; liver and kidney function; muscle tissue breakdown; and dietary carbohydrate amount and rumen degradability. Creatinine is used to assess renal function and its impact on BUN values. Total protein and albumin reflect availability of protein and their concentration decline in the face of protein deficiency. However, this occurs over a period of time. Albumin has a relatively short half-life and can reflect protein deficiency problems over a period of a month or two. Creatine kinase is released from muscle when it is catabolized or injured. In most dietary protein deficiency situations, BUN values will be low (<10 mg/dl [<7.1 mmol/l]) with normal albumin concentration (>3.5 g/dl) in the early dry cows. Close-up dry cows will have low to moderate BUN, lower albumin and elevated Ck values. Fresh cows generally have low BUN and low albumin (<2.5 g/dl). These fresh cows seemingly fail to properly respond to any disease insult. Protein deficient fresh cows will die from metritis, mastitis, foot rot, and anything else without antibiotic therapy. An interpretation of this situation is that there are no amino acids available to support the immune system and it fails, predisposing the animal to any bug that comes along.

8.2.3. Assessment of micronutrient status

Macrominerals calcium (Ca), phosphorus (P), potassium (K), magnesium (Mg), sodium (Na), chloride (Cl), and sulfur (S) are of extreme interest as to their status relative to their role in milk fever, alert downer cows, and weak cow syndrome. Unfortunately, most of these minerals are tightly regulated in the body through a variety of homeostatic processes. Blood concentrations of macrominerals are not reflective of dietary status when the homeostatic system is functioning properly. Phosphorus, K, Mg, and S are macrominerals in which blood concentrations are somewhat sensitive to dietary intake. Sodium and chloride concentrations are altered when renal or digestive function is compromised or in extreme dietary deficiency states. Assessment of Ca concentrations around the time of calving is a useful indicator of how well the Ca regulatory system is working and potential for clinical or subclinical hypocalcemia problems "Other than the 2 weeks prior to and following calving, blood Ca is not a very sensitive diagnostic measure as a result of the intact regulatory system. Therefore macromineral blood concentrations will need to be carefully interpreted in light of whether or not the homeostatic system is in proper operation. In lieu of directly measuring macromineral blood concentration, measurements of parameters directly related to functioning and responsiveness of the homeostatic regulatory system may offer some insight as to nutritional status. Other methods of determining mineral balance such as urinary excretion patterns are being investigated as more sensitive indicators of nutritional status.

Assessment of trace mineral and fat-soluble vitamin status is routinely completed using direct blood concentration measurements. The question to ask is whether or not there is a predictive relationship between tissue or blood trace mineral or fat-soluble vitamin concentrations and presence of nutrient-specific deficiency disease. On the surface one would have to say yes because we can document low nutrient concentrations in the presence of disease signs. The question really becomes one of how predictive mineral and vitamin concentrations are and which ones are the best indicators. To understand the issue here we need to appreciate that trace minerals and fat-soluble vitamins are not in a single large pool in the body, but are distributed into a number of different pools which have different functions and availability. The different nutrient pools described include a storage, transport, and biochemical function pools. As a result of the storage capacity for trace minerals and fat-soluble vitamins in the liver, moderate dietary deficiencies or short-term severe deficiencies can be overcome without any effect on the critical biochemical functions performed by the element in question. If the dietary insult is severe or prolonged enough to drain the storage pool, then some effects might be seen in the transport pool. Finally when the transport pool has been compromised, the biochemical function pool will be compromised resulting in some dysfunction. It is only when the biochemical function pool reaches a critically low level that we see the overt clinical deficiency disease. However, before we reach the clinical disease stage, we will see problems associated with subclinical disease including increased disease susceptibility as a result of compromised immune function. Therefore, the bigger challenge is to pinpoint the ability of the chosen marker to be measured and its relationship to changes in one or more of these trace mineral or fat-soluble vitamin pools. Most of our markers currently being used are element concentrations in either whole blood or serum that probably reflect the transport pool and not necessarily the biochemical function pool. As a result they may not be highly correlated with the presence of clinical disease.

e.g.

i) *serum Cu concentrations*: Unless serum Cu is a critically low, it has no significant predictive value in assessing potential for Cu deficiency disease.

ii) serum and whole blood Se concentration: Serum Se values represent the transport pool and are very sensitive to dietary changes and liver mobilization, while whole blood Se values represent both transport and a portion of the biochemical function pools. This measure is somewhat less sensitive to dietary changes due to greater proportion of whole blood Se being present as the erythrocyte enzyme glutathione peroxidase. If one was to assess a potential response to a dietary change, serum Se values would respond within a day or two while whole blood may take a month or more to show a significant change.

Liver mineral concentrations are good markers for the storage pool; however, they are not always highly associated with the presence of disease. Liver mineral concentrations may give us some insight into the adequacy of the mineral program and potential for disease. We can also go for assessment of mineral status in fetal and neonatal animals. Research has shown that the fetus can concentrate trace minerals in its liver and therefore, comparison to adult values is inappropriate. Secondly, fetal liver has a lower dry matter content than the maternal liver further substantiating the inability for direct comparison. However, databases determining normal trace mineral concentrations in the fetal and neonatal liver need to be developed.

8.2.4. Liver function evaluation

Liver function can be assessed through a variety of enzymes alanine aminotransaminase-ALT (gamma-glutamyl transferase-GGT or glumate puruvate transaminase-GPT), aspartate aminotransferase-AST (glumate oxaloacetate transaminase-GOT) and sorbitol dehydrogenase-SDH and total bilirubin concentrations in the blood. Unfortunately, an elevation in any of these parameters does not mean anything more than some insult has occurred to the liver. Bilirubin values are most specific to bile flow problems than overt liver cell damage. These enzyme values need to be interpreted in conjunction with total cholesterol and NEFA results. As described for energy balance, NEFAs are released into the circulation as a direct result of fat breakdown. The liver takes up NEFA in direct relationship with their concentration in blood. Once in the liver, NEFAs can either be partially metabolized to ketone bodies and distributed to other tissues for energy metabolism or they can be used to synthesize fat. High NEFA values result in either elevated ketones or fat production by the liver. Fat in the liver has two potential options, remain in the liver cell and initiate hepatic lipidosis (fatty liver) or be transported out of the liver. The lipoprotein structure that transports fat from the liver is identified as a very low density lipoprotein (VLDL). Associated with fat in the VLDL structure is a substantial amount of cholesterol. Therefore, total serum cholesterol indirectly measures the presence of VLDL in blood and consequently measures the liver's ability to produce VLDL. If VLDL production is compromised, hepatic fatty infiltration will ensue. Therefore the values described above represent total cholesterol values that characterize conditions in which VLDL production is limited and fatty infiltration is probable. Some investigators have suggested assessing the NEFA to Cholesterol ratio for this reason.

8.2.5. Metaboiites diagnostic to nutritional diseases

For individual animals, metabolite values are compared to standard, laboratory-dependent reference values. These reference values generally represent a 95% confidence interval. This means that 95% of normal animals should have a given metabolite concentration within this range. This also suggests that 5% of the population will be outside of this reference range and still be normal, emphasizing the need to clinically evaluate the animal. A number of factors, most notably physiologic state and age have been shown to influence blood metabolite concentrations. Most reference ranges do not account for these differences and thus may confound direct interpretation. Having a thorough understanding of the physiologic regulation of a given nutrient is crucial to interpretation. For trace minerals, blood or serum concentrations are buffered from acute changes as a result of dietary problems through mobilization of storage mineral, usually from the liver. This suggests that liver trace mineral status may be a better indicator of dietary adequacy, whereas measurement of a mineral-specific enzyme activity better reflects the presence of overt clinical deficiency disease compared to blood concentrations. Many trace mineral concentrations in blood are influenced by disease. As we come to better understand the factors that affect metabolites, we can adjust and better assess nutritional status. In contrast to individual animal samples, pooled mean metabolite values cannot be directly compared to reference ranges in the same way. When interpreting pooled samples one needs to remember that measured value represents a population with individuals above and below the mean. As a general rule, means of pooled samples should be near the midpoint of the reference range to be considered normal. For example, if serum total calcium (Ca) concentration for fresh cows is 9 mg/dl (2.25 mmol/L) and the reference range is 9 to 12 mg/dL (2.25 to 2.99 mmol/L), this might be interpreted to suggest a potential problem with subclinical hypocalcemia whereas it would be considered normal in an individual. The measured mean of 9 mg/dl (2.25 mmol/L) represents a population with approximately 50% of the individual values above and below. This suggests that a number of individuals would have serum Ca concentrations below the normal range. Of course interpretation of metabolic profile results has to be considered in light of presenting problems in the herd. If the herd is experiencing clinical signs consistent with subclinical hypocalcemia, e.g., slow increase in feed intake and milk production, displaced abomasum and ketosis problems, this would be supportive evidence of the metabolic profile results. For low BUN values, it is difficult to have values approaching zero whereas for other metabolites, if the sampled cow had an extremely skewed value, it would have been exhibiting clinical signs and would not have been sampled. Metabolites with high variability (wide range of values) will be of less diagnostic value as compared to low variability metabolites. Various conditions associated with increase or decrease in blood metabolite concentration are tabulated below.

Metabolites	Condition	Associated factors
Glucose	Hypoglycemia	Liver dysfunction, chronic starvation, malabsorption, ketosis, pancreatic tumour, hyperinsulinism.
	Hyperglycemia	Diabetes mellitus, convulsions, following a meal, insulin overdose, pancreatitis, fear, excitement.
Urea	Increase	High protein or urea feeding, carbohydrate deficiency, catabolic drugs or increased catabolism, dehydration, severe haemorrhage, fever and necrosis, prolonged exercise, hyperthyroidism.
	Decrease	Low protein diet, diabetes insipidus, liver dysfunction, anabolic steroids, malnutrition, aflatoxicosis.
Creatinine	Increase	Dehydration, primary and chronic renal failure, severe exercise, high

		NPN feeding, muscular degeneration, lipemia, hemolysis, ketosis.
Protein	Hyperproteinaemia	Dehydration, acute, subacute and chronic inflamation, liver disease, neoplasia, autoimmune disorders, increased globulin, newborn, haemolysis, glomerular disease, anabolic steroids, uremia.
	Hypoproteinaemia	Protein starvation, liver disorders, intestinal malabsorption, haemorrhage, burns, congestive heart failure, wasting disease.
Albumin: Globulin ratio	Increased	Hypogammaglobulinaemia, congenital and acquired immuno-deficiency, uraemia.
	Decreased	Increased globulin, decreased albumin; anaphylaxis, following imm uniza tion, protein starvation, hypoproteiriaemia.
Bilirubin	Hyperbilirubinaemia	Severe acute haemolysis, liver dysfunction, cholestasis.
Cholesterol	Hypercholesterolaemia	Following a fatty meal, hypothyroidism, diabetes melitus, starvation, severe trauma, liver dysfunction, nephrotic syndrome, steatitis, cholestasis.
	Hypocholesterolaemia	Hepatitis, hyperthyroidism, low fat diets, malnutrition, severe anaemia, intestinal malabsorption.
Calcium	Hypercalcemia	Hyperparathyroidism, hypervitaminosis D, osteoneop lastic disease, high calcium diet
	Hypocalcemia	Hypoparathyroidism, milk fever, osteoporosis, steatorrhoea, nephritis, hypovitaminosis D, low calcium diet, ricket
Phosphorous	Hyperphosphataemia	Chronic nephritis, hypoparathyroidism, hypervitaminosis D, high P diet, wheat poisoning

	Hypophosphataemia	Rickets, hyperparathyroidism, pica, low P diet, periparturient haemoglobinuria, osteoporosis, secondary hyperparathyroidism.
Alkaline phosphatase	Increase	Bone disease, severe starvation, increased osteoplastic activity, increased bone metabolism, renal diseases, hepatobiliary diseases, osteomalacia, rickets, cholestasis, active rowth, last trimester of pregnancy.
Glutamic puruvate transminase (GPT/ALT)	Increase	Hepatitis, obstructive jundice, fever, acute pancreatitis, muscular damage
Glutamic oxaloacetate transaminase (GOT/AST)	Increase	Liver damage, muscular damage, severe exertion
Lactate dehydrogenase	Increase	Liver disorder, muscular damage
Creatine phosphokinase	Increase	muscular damage, severe exertion, hypoparathyroidism, CNS disorder.
Sorbitol dehydrogenase	Increase	Liver damage.

8.2.6. Collection and preservation of blood samples

Time of sample collection

Fixing a time schedule is important in eliminating the possible effects of diurnal variations and the suitable time is usually in the morning (8.00 10.00) after an overnight fasting.

Sites of collection

Animal species	**Site of collection**
Cattle, Buffalo, Horse, Sheep, Goat	Jugular and Femoral vein
Swine	Ear vein, Anterior venacava
Dog	Cephalic vein, External saphenous vein
Cat, Rabbit,	Ear vein
Chicken	Wing vein on inner side of elbow joint
Rat, Mice	Heart puncture, Tail sniping
Guinea pig	Heart puncture

Commonly used anticoagulants

Anticoagulant	mg/ml blood	Use	Mode of action
Heparin	0.2	Procedures requiring whole blood or plasma, especially useful when intact erythrocytes are desired	Inhibits conversion of prothrombin to thrombin.
Ethylenediamine tetra-acetic acid (Disodium salt)	1.0	Procedures requiring whole blood or plasma	Chelates ionic calcium (Ca^{++}).
Oxalates of Lithium Sodium Potassium Ammonium	1.0-2.0	Procedures requiring whole blood or plasma; causes shrinkage of cell volume. Should not be used in hematological procedures.	Forms un-ionized Ca oxalate complex; reduces level of ionized Ca below that required for clotting.
Sodium citrate	5.0	Similar to oxalates	Forms un-ionized complex with Ca,
Sodium fluoride	10.0	Both anticoagulant and preservative for blood glucose determinations by inhibiting blood enzymes causing glycolysis; causes erythrocyte shrinkage. Combined with thymol (10: 1) for effective control of microbial growth in stored blood samples.	Forms un-ionized calcium fluoride complex.

Collection of blood

(i) The needle and syringe must be dry (water hemolyses RBCs)

(ii) Clip the area from where blood is to be drawn

(iii) Apply tincture of iodine and let it to dry

(iv) Raise the vein by pressure

(v) Insert the needle into the vein and draw required quantity of blood

(vi) Disengage the needle, and collect the blood into the specimen tube (with or without anticoagulant)

(vii) Mix blood with anticoagulant by rotatory movement. Never shake which produces froth and RBCs may be ruptured.

Preparation of blood serum

Place the freshly drawn blood directly into a clean dry test tube without anticoagulant. Allow the blood to clot at room temperature. Separate the clot from the walls of the tube by carefully running a clean applicator stick around the inner surface of the tube and centrifuge immediately (Preferebly in a refrigerated centrifuge) and collect the supernatant serum. Preserve it at -20°C for further analysis.

Preparation of blood plasma

Freshly drawn blood is collected in a clean dry test tube with an anticoagulant and then centrifuged to separate out the supernatant as plasma. Preserve it at -20°C for further analysis.

8.3. Estimation of Blood Biochemical Constituents

8.3.1. Total protein

Biuret method

Principle

The protein is measured by the intensity of purple colour which results by reaction of -CONH group of protein molecule with copper sulphate in alkaline medium.

Reagents

1. *Biuret reagent*: Take 0.75g CuS04 and 3.0g sodium potassium tartrate in 500ml. volumetric flask. Add 250 ml. distilled water to dissolve the above solids. Add 150ml. 10% (w/v) NaOH while stirring the contents of the flask vigorously. Make up the volume to 500 ml with distilled water. The reagent can be stored for about 6 months at room temperature. If a black precipitate is observed, the solution should be discarded.

2. *Standard bovine serum albumin solution (40mg/ml):* Dissolve 400mg BSA in distilled water and make up the volume to 10 ml.

Procedure

1. Prepare tubes in duplicate as follows:

Tube No.	BSA Standard solution (ml)	Distilled water (ml)	BSA concentration (mg)
Blank	0.0	1.0	0.0
1	0.1	0.9	4
2	0.2	0.8	8
3	0.3	0.7	12
4	0.4	0.6	16
5	0.5	0.5	20
6	0.6	0.4	24
7	0.7	0.3	28
8	0.8	0.2	32
9	1.0	0.0	40

2. Add 4.0 ml Biuret reagent to each tube and vortex the mixture.
3. Incubate the tubes for 20 minutes at 37°C.
4. Take absorbence at 540nm against the black. Colour is stable only for 1 hour.
5. Plot the standard curve of BSA.
6. Take 0.5 ml. serum sample and proceed in the same way and determine their protein contents from the standard curve.

Reference

Hiller, McIntosh and Van Styke 1927. Colorimetric determination of proteins. J. Clin. Invest. 4: 235.

8.3.2. Albumin and globulins

Principle

At pH 4.2, serum albumin binds with bromocresol green and the colour intensity is directly proportional to the concentration of albuminGlobulin is simply the substraction of albumin from total protein in serum.

Reagents

1. Succinate buffer (0.1 M, pH 4.0): 5.95g succinic acid is dissolved in 400 ml H_20 and pH is adjusted to 4.0 with O.lN NaOH and volume is made to 500 ml.
2. Bromocresol green solution (stock) 0.6 mM: Dissolve 41.9 mg BCG in l' ml. O.lN NaOH and dilute to 100 ml, H_2O
3. Working dye solution: 1 vol. of reagent 2 and 3 vol of reagent 1 is mixed. Brij-35 (30% w/v) solution is added at 0.4% (v/v) and the pH is adjusted to 4.2±0.05
4. Standard albumin solution (40mg/ml): Take 400mg bovine serum albumin and dissolve in 10 ml normal saline.

Procedure

1. The estimation procedure follows the protocol given below:

Protocol

Reagents/Sarnpletul)	Blank	Sample	Standard				
			1	2	3	4	5
Normal Saline (Ill)	50.0	25.0	0.0	10.0	20.0	30.0	40.0
Standard albumin (ug)	-	-	50.0	40.0	30.0	20.0	10.0
Sample	-	25.0	-	-	-	-	-
Reagent 3 (ml)	3.0	3.0	3.0	3.0	3.0	3.0	3.0

2. Mix well and stand for 10 minutes at room temperature
3. Read OD at 630 nm against blank
4. Plot a satandard curve and read the concentration of sample.

   N.B. Ammonium sulphate precipitation at different concentration also separates fibrinogen, globulins and albumins and the fraction can be estimated by biurete method.

Reference

Doumas, B.T., Watson, W.A. and Biggs, H.G. 1971. Albumin standards and the measurement of serum albumin with Bromocresol green. Clin. Chim. Acta. 31: 87-96.

8.3.3. Glucose

Principle

Glucose is reduced in the presence of O-toluidine solution to emit a bluish green colour which can be measured at 625nm.

Reagents

1. Trichloroacetic acid solution (3%, w/v): Take 3g TCA and dissolve in 100 ml. distilled water.
2. O-toluidine solution(6%,v/v): Mix 6 ml. O-toluidine with 94 m!. glacial acetic acid.
3. Glucose standard solution = Take 120 mg glucose and dissolve in 100 ml distilled water.

Procedure

1. Take 0.1 ml. blood in a test tube and add to it 1.0 ml TCA for protein precipitation.
2. Centrifuge the mixture at 2000rpm for 5 min. and collect the supernatant.
3. Prepare the tubes in duplicate according to the protocol given below

Reagents/Solutions (ml.)	**Blank**	**Sample**	**Standard**				
			1	**2**	**3**	**4**	**5**
Protein free sample	-	0.20	-	-	-	-	-
Standard	-	-	0.04	0.08	0.12	0.16	0.20
Distilled water	0.2	-	0.16	0.12	0.08	0.04	0.00
O-toluidine reagent	3.00	3.00	3.00	3.00	3.00	3.00	3.00

4. Boil in water bath for 8 min and then cool under tap water.
5. Absorbance is read at 625nm against reagent blank.
6. Plot a standard curve and read the concentration of unknown sample.

Calculation

$$\text{Glucose (mg/100ml)} = \frac{\text{OD unknown}}{\text{OD s tan dard}} \times \text{conc. of standard} \times \text{dilution factor}$$

Note

1. For estimation of glucose in serum/plasma omit step no. 1 and 2. However, if sample is haemolysed, step no. 1 and 2 are necessary.
2. Glucose estimation should be done immediately after sample collection.
3. Benzoic acid does not interfere with reaction and can safely be used as preservative and Glucose standard solution can be stored for 2 weeks by adding a pinch of benzoic acid.
4. The reaction is very sensitive to water and so the dilution should not exceed 10 times.
5. For plasma sample a mixture (10:1) of sodium fluoride and potassium oxalate should be used at the rate of 1 mg/ml as anticoagulant.

Reference

Hultmann, E. 1959. Rapid specific method for determination of aldohexoses (aldosaccharides) in body fluids. Nature. 103: 108 - 109.

8.3.4. Urea

Principle

A rose pink colour complex is formed when urea is treated with diacetyl monoxime and thiosemicarbazide in the presence of sulphuric acid, phosphoric acid and ferric chloride which is read at 525nm in a spectrophotometer.

Reagents

1. Reagent A: 300 ml conc. H_2SO_4 (95-98%)
   50 ml. distilled water
   100 ml. conc. H_3PO_4 (85%)
   100 mg Ferric chloride ($FeCl_3$)
   Mix well and volume it to 1L with distilled water.
2. Reagent B: 500mg Diacetyl monoxime (DAMo)
   10 mg Thiosemicarbazide (TSC)
   Mix well with distilled water to make 100ml solution
3. Reagent C: Prepare immediately before use by mixing reagents A and B at 2: 1 proportion.

4. Trichloroacetic acid (TCA) 5% (w/v): Weigh 5g TCA and dissolve in 100 ml distilled water
5. Stock standard urea solution (500 /lg/ml): Weigh 50mg urea and dissolve in 100 ml distilled water.
6. Working standard (50/lg/ml): Dilute 1 vol. of stock with 9 vol. of distilled water.

Procedure

1. Protein is precipitated out from serum/plasma by mixing 0.2 ml. with 1.8 ml of 5% TCA.
2. It is centrifuged at 2000 rpm for 5 min and supernatant is collected for further processing using the protocol given below.

Reagents/Solutions (ml.)	Blank	Sample	Standard				
			1	2	3	4	5
Protein free sample	0.20	-	-	-	-	-	-
Standard	-	-	0.02	0.04	0.06	0.08	0.10
Distilled water	-	0.20	0.18	0.16	0.14	0.12	0.10
Reagent C	3.00	3.00	3.00	3.00	3.00	3.00	3.00

3. The tubes are then incubated in boiling water bath for 5 minutes, exactly.
4. Cool to room temperature and absorbance is read at 525nm against reagent blank.
5. Plot a standard curve and read concentration of the sample against absorbance obtained.

Calculations

Urea (mg/100 ml) = OD unknown × conc. of standard × dilution factor

OD standard

Reference

Rahmatullah, M. and Boyde, T.R.C. 1980. An improvement in determination of urea using diacetylmonoxine method with and without deproteinization. Clin. Chern. Acta. 107: 3 - 9.

8.3.5. Total cholesterol

Principle

Cholestrol is oxidised by FeCb-H2S04 to yield purple colour which is measured spectrophotometrically at 560 nm.

Reagents

1. Glacial acetic acid (aldehyde free)
2. Ferric chloride (FeCls) solution (10%, w/v) weigh Ig ferric chloride and dissolve in 10 ml. glacial acetic acid.
3. Colour Reagent (should be prepared fresh) 1 ml. reagent 2 is mixed well with 15 ml. cone. H2S04 and then volume is made to 100 ml. with cone. H2S04.
4. Standard cholesterol solution (0.25%, w/v). Weigh 25mg cholesterol and dissolve in 10 ml. glacial acetic acid.

Procedure

1. Sample and standard are run according to the following protocol

Protocol

Reagents/Solutions (ml.)	Blank	Sample	Standard				
			1	2	3	4	5
Reagent 1	3.00	3.00	3.00	3.00	3.00	3.00	3.00
Distilled Water	0.1	0.00	0.08	0.06	0.04	0.02	0.00
Sample	-	0.1	-	-	-	-	-
Standard Cholesterol	-	-	0.02	0.04	0.06	0.08	0.10
Reagent 3	2.00	2.00	2.00	2.00	2.00	2.00	2.00

2. The whole content is mixed vigorously and kept for 20 min. at room temperature and absorbance is read at 570nm against the blank.
3. Plot a standard curve taking absorbance against concentration of standard cholestrol and work out for the unknown sample.

Calculation

Serum/plasma cholesterol (mg/100ml) = OD Sample X cone. of standard OD standard

Note: The colour reagent (3) is added carefully down the tube to form a layer under acetic acid.

Reference

Zlatkis, A., Zak, B., Boyle, H.J.a nd Mich, D. 1953. A new method for direct determination of serum cholesterol. J. Lab. Clin. Med 41: 486 - 492.

8.3.6. Creatinine

Principle

Creatinine reacts with alkaline picrate to form a yellow-red colour known as the Taffe reaction.

Reagents

1. Sodium tungstate (10%, w/v): Weigh 109 sodium tungstate and dissolve in distilled water to make volume 100 ml.
2. Sulphuric acid (0.66 N)
3. Sodium Hydroxide (10%, w/v) Weigh 109 sodium hydroxide crystals and dissolve in distilled water, cool and make volume to 100 ml.
4. Picric acid saturated solution in H20.
5. Hydrochloric acid solution (0. IN)
6. Alkaline picrate reagent. Reagent 3 and 4 is mixed at 1:5 ratio.
7. Standard creatinine solution

   Stock (1mg/ml): Dissolve 0.1 g creatinine in 100 ml. O.IN HCl.

   Working standard (10ìg/ml): 1 ml stock is mixed with 10 ml. 0.1 N HCl and the volume is made to 100 ml. with H_2O.

Procedure

1. Take 0.5 ml serum/plasma in a test tube, add 3.5 ml H20 and 0.5 ml reagent 2. Shake well and stand for 10 minutes.
2. Add 0.5 ml reagent 1, mix well, stand for 10 min. and then centrifuge at 3000 rpm.
3. Filtrate out the supernatant and run sample blank, filtrate and standard as per the protocol given below.

Protocol

Sample/Reagent (ml)	Blank	Sample	Standard				
			1	2	3	4	5
H_2O	2.5	-	2.0	1.5	1.0	0.5	0.0
Standard	-	-	0.5	1.0	1.5	2.0	2.5
Conc. of standard (ìg)	-	-	5.0	10.0	15.0	20.0	25.0
Sample filtrate	-	2.5	-	-	-	-	-
Reagent 6	2.5	2.5	2.5	2.5	2.5	2.5	2.5

4. Mix well and stand at room temperature for 20 min.
5. Absorbance is read in spectrophotometer at 525nm against the blank.

Calculation

$$\text{Creatinine (mg/100ml)} = \frac{\text{OD sample}}{\text{OD standard}} \times \text{conc. of standard} \times \text{dilution factor}$$

Reference

Metate, R.D. and Levinson, S.A. 1954. Am. J. Clin. Path. 24: 511 (c.f. clinical and Laboratory Diagnosis by Levinson, S.A. and Mefate, R.D. 91969). 7th ed. Lead and Febiger, Philadelphia, pp 413 - 405.

8.3.7. Bilirubin

Principle

The coupling of bilirubin with O-benzenediazonium sulfonate to form ozobilirubin is used for the estimation. Photometric estimation of the purple colour developed exactly 5 minutes after diazotisation with sulfanilic acid in aquous solution is a measure of direct bilirubin. The colour developed after adding methanol at a level that releases bound bilirubin into solution without precipitating proteins represents total bilirubin.

Reagents

1. *Sulfanilic acid solution*: Dissolve 100 mg of sulfanilic acid in 1.5 ml conc. hydrochloric acid and add water to make 100 ml.
2. 25% sodium nitrite stock solution in water. Store at 4°C. Prepare working solution just before use by diluting stock sodium nitrite solution 1:50 with water (0.5%).

3. Diazo reagent: Prepare just before use by mixing 0.3 ml of 0.5% sodium nitrite solution and 10 ml, sulfanilic acid solution.
4. Diazo blank solution: Dilute 0.5m!. cone. HCI to 100 ml. with water.

Procedure

1. Prepare in duplicate by taking 0.4 ml. plasma and 3.6 ml. water in 'test' and 4.0 ml water in blank test tubes.
2. Add 1.0 ml diazo blank solution in blank tubes and 1.0 ml. diazo reagent in other test tubes and mix immediately.
3. Read absorbance (E1) of test exactly 5 minutes after addition of diazo reagent at 540nm against blank.
4. Add 5.0m!. absolute methanol to each tube and mix by gentle inversion.
5. Leave for 30 minutes and read absorbance (E2) at 540 nm by avoiding formation of air bubble in cuvette.

Calculation

The absorbance of azobilirubin equivalent to 5mg bilirubin/lOOml at 540 nm is 0.211;

Direct bilirubin (mgllOO ml plasma) = E1 × 11.85

Total bilirubin (mg/IOO ml plasma) = E2 × 23.70

8.3.8. Ascorbic acid

Principle

Method for the estimation of Ascorbic acid is based on the measurement of colour of the product formed by the coupling of 2,4 dinitrophenylhydrazine with the ketone group of oxidised ascorbic acid.

Reagents

1. *2-4-dinitrophenylhydrazine*: 2g 2,4-dinitro-phenylhydrazine (DNPH) IS dissolved to make 100m I. with approximately ION H2S04 and filtered.
2. *Trichloro acetic acid. (TCA):* 6g TCA is dissolved in distilled water to make volume 100 ml.

3. Acid washed norit (charcoal): 200 g of norit is suspended in 1 litre of 10% HCL Keep for boiling and filter under suction. The cake is removed and stirred with 1 litre of water and refiltered. This procedure is repeated until washings give a negative test for ferric ions. The norit is then dned overnight at 110-120°C.
4. Thiourea: 109 of thiourea dissolved in 10 ml. of 50% aquous alcohol.
5. Standard ascorbic acid. Take 100mg ascorbic acid and make volume 100ml. with 6% TCA solution.

Procedure

1. Take 2ml plasma and 6ml 6% TCA in centrifuge tube: After 5 min centrifuge at 2000 rpm for 5 min. and collect the supernatant.
2. Add 200 mg of acid washed norit in supernatant and shake vigorously.
3. Take 1.6 ml of norit treated supernatant and add to it one drop of thiourea solution and 0.4 ml of DNPH reagent.
4. Prepare sample blank tube containin].6 ml. norit treated supernatant and a drop of thiourea
5. The tubes are incubated at 37°C for 3h and then placed in a ice water bath.
6. Colour is developed by adding drop wise 2 ml. cone. H2S04 with continuous stirring.
7. Add 0.4 ml. DNPH reagent in the sample blank tube.
8. After 30 minutes measure OD of sample at 540 nm against sample blank.
9. Prepare the standard tubes in duplicate as follows

Tube No.	Blank	Standard								
		1	2	3	4	5	6	7	8	9
6% TCA	1.6	1.5	1.4	1.3	1.2	1.0	0.8	0.6	0.2	0.0
Std. Ascrobic acid	-	0.1	0.2	0.3	0.0	0.6	0.8	1.0	1.4	1.6
Conc. of Ascrobic acid	0.0	0.1	0.2	0.3	0.4	0.6	0.8	1.0	1.4	1.6

10. Proceed for colour development following steps No.3, 5 and 6.
11. Read absorbance at 540 nm against the blank.

12. Read OD of sample on the standard curve and calculate the concentration of ascorbic acid.

Note: Ascorbic acid estimation should be done immediately after collection of sample. Use of phsophotungstic acid is also recommended for its estimation.

Reference

Pijoan, M. and Gerjovich, H. J. 1946. The use of 2,4-dinitrophenylhydrazine for the determination of ascorbic acid. Science 103 (2668), 202-203.

Kyaw, A. 1978. A simple colorimetric method for ascorbic acid determination in blood plasma. Clin Chim Acta. 86:153-157.

8.3.9. Pyruvic acid

Principle

Deproteinised serum is treated with 2,4-dinitrophenylhydrazine forming a 2,4-dinitrophenylhydrazone which reacts with strong alkali to form a redish compound that is estimated photometerically.

Reagents

1. Trichloroacetic acid 10%: Take 109 TCA and dissolve in distilled water to make the volume to 100 ml.
2. 2,4-dinitrophenylhydrazine(DNPH)0.1%: 100 mg DNPH in 100 ml distilled water.
3. Sodium carbonate 10%: Dissolve 109 sodium carbonate in distilled water and make the volume to 100 ml.
4. Sodium hydroxide 1.5 N
5. Stock standard pyruvic acid (lmg/ml): Take 100 mg pyruvic acid and dissolve in distilled water to make the volume 100 ml.
6. Working standard pyruvic acidf Iflug/rnl): Take 1 ml. stock solution of pyruvic acid in 100 ml volumetric flask and make the volume with distilled water.
7. Xylene, toluene or benzene.

Procedure

1. Take 1.0 ml freshly drawn blood in a centrifuge tube, kept in ice containing 5.0 ml 10% TCA and centrifuge at 2,000 rpm for 5 min. at 4°C.

2. Pip pet 1.5 ml. of the supernatant into a test tube and allow to come at room temperature
3. For standard take 1.0 ml working standard pyruvic acid solution and 0.5 ml distilled water.
4. Add 0.5 ml of DNPH and allow to react at room temperature for exactly 5 mm.
5. To each tube add 1.5 ml xylene/benzene/toluene and pass a stream of air through the mixture for 2 min.
6. After settling remove the lower layer by means of dropper with a capillary tip.
7. Add 3 ml. 10% sod. carbonate to each tube and again mix by bubbling air for 2 min.
8. After settling the mixture pipet 2.5 ml. of aqueous (lower) layer and place in the photometer cuvet.
9. Add 2.5 ml 1.5 N NaOH, mix and leave for 5-10 min. and read OD at 520nm by using blank (a mixture of 2.5ml 10% sod. carbonate and 2.5 ml 1.5N sodium hydroxide).

Calculations

Pyruvic Acid (mg/lOO ml) = Density of unknown X Cone. of standard Density of standard.

Reference

Katsuki H., Kawano C., Yoshida T., Kanayuki, H. and Tanaka S. 1961. The determination of pyruvic acid by 2,4-dinitrophenylhydrazine method.Analytical Biochemistry, 2, 433-440.

8.3.10. Calcium

Principle

Serum calcium is precipitated as oxalate and is titrated with standard potassium permanganate.

Reagents

1. Ammonium oxalate solution (4%, w/v), 4g ammonium oxalate is dissolved in distilled water to make volume 100 ml.
2. Ammonia solution. Liquor ammonia is diluted to 2% with istilled water.

3. IN H_2SO_4
4. O.01N potassium permanganate solution

Procedure

1. Two ml of serum is taken in 15ml centrifuge tube. Add 2 ml water and 1 ml 4% ammonium oxalate solution and mix thoroughly. Leave for 30 min. Mix and centrifuge at 1500 rpm for 5 min.
2. Carefully pour off the supernatant by inverting the tube for 5 min on a rack. The mouth of the centrifuge tube is wiped with a piece of filter paper.
3. Wash the sides of the tube with 3 ml dilute ammonia solution and stir the precipitate. Centrifuge the tube and drain off the supernatant again as before.
4. Wash the precipitate once more with dilute ammonia to ensure complete removal of ammonium oxalate.
5. Add 2 ml 1N H_2SO_4 to dissolve the precipitate.
6. Place the tube in boiling water bath for 1 minute.
7. Titrate with O.OIN KMn04 solution to a definite pink colour persisting for at least one minute (A).
8. Run blank with 2 ml of IN H_2SO_4 in a tube (B).

Calculation

1ml of O.OlN KMn04 is equivalent to 0.2004 mg of calcium. Calcium (mg/100 ml serum) = A - B X 0.2004 X 100/2

Reference

Clark, E.P. and Collip, J.B. (1925). Determination of serum calcium. J. Biol. Chem. 63:641.

5.3.11. Phosphorus

Principle

When protein free filtrate is treated with acid molybdate solution, the phosphate present in the filtrate forms phosphomolybdic acid. The phosphomolybdic acid is reduced by 1, 2, 4- amino naphthol sulfonic acid to develop a blue colour which is measured colorimetrically.

Reagents

1. 10% trichloroacetic acid (TCA). Weigh 109 TCA and dissolve in distilled water to make volume 100 mi.
2. Molybdate solution: Put 2.5g ammonium molybdate in 20 ml water. Add 30 ml of ION H2S04 with periodic shaking until dissolved. Make the volume 100 ml with water.
3. Aminonaphthol sulfonic acid reagent: Take 195 ml of 15% sodium bisulphite in a cylinder. Add 0.5 g 1,2,4 amino napthol sulphonic acid and 5 ml of 20% sodium sulphite. Shake the contents and store in brown bottle.
4. Standard phosphate solution (0.8 mg/ml): Stock: 351 mg dry KH2P04 in 10 mILO N H2S04 and make the volume 100 ml with distilled water.
5. Working standard solution (Sug/rnl.): Take 1.0 ml. of stock standard solution and dilute with distilled water to make the volume 100ml.

Procedure

1. Take 8 ml 10% TCA in a small flask and add slowly with stirring 2 ml blood, plasma or serum and shake well.
2. Filter through filter paper No. 40. 41 or 42 and collect the filtrate.
3. Prepare tubes in duplicate as follows.

Tube No.	**Blank**	**Sample**	**Standard**				
	1	**2**	**3**	**4**	**5**	**6**	**7**
Distilled water (ml)	5.0	-	2.0	1.5	1.0	0.5	0.0
Working standard (ml)	-	-	3.0	3.5	4.0	4.5	5.0
Sample filtrate (ml)	-	5.0	-	-	-	-	-
Reagent 2 (ml)	1.0	1.0	1.0	1.0	1.0	1.0	1.0
Reagent 3 (ml)	0.4	0.4	0.4	0.4	0.4	0.4	0.4
Distilled water (ml)	3.6	3.6	3.6	3.6	3.6	3.6	3.6

4. Mix and allow to stand for 5 min.
5. Read absorbance of standard and sample at 700nm against the blank.

Calculation

$$\text{Phosphorus (mg/100 ml)} = \frac{\text{OD unknown}}{\text{OD of known}} \times 0.04 \times 100$$

Reference

Fiske, C.H. and Subbarow, Y. (1925). The colorimetric determination of phosphorous. J. Biol. Chern. 66: 375.

8.3. Free fatty acids

Equipment

12 mm × 100 mm screw-cap culture tubes with teflon-lined caps, Shaker, Centrifuge, Spectrophotometer.

Reagents

1. Redistilled chloroform - keep refrigerated in brown bottle - use within 7 d.
2. *Chloroform*: Heptane:Methanol- 4:3:2 ratio - make up on the day of analysis.
3. .035 N HCl- 3 mL ofHCl diluted to 1 L with water.
4. Stock salt solution - 200 g ofNa2S04 + 100 g of Li_2SO_4 + 40 g of CO (N03)2'6 H_2O - make to 1 L with water - can be stored indefinitely.
5. Salt reagent - make up day of analysis - add 8 mL of triethanolamine to 100 mL of stock salt solution - let stand 15 min before use.
6. Indicator solution -.2 g of I-nitroso-2-naphthol in 1 L of 95% ethanol. Store in brown bottle at room temperature.
7. Standards: Blank = heptane

   Stock solution - 2,000 ìmoles of palmitic acid/L = 0.5128 g of palmitic acid/L of heptane

   Working Standards: Dilute stock solution with heptane to yield 1,000, 500, and 250 µmoles/L

Standards and chemicals

1. CHCb - Wash with 5% (vol/vol) H_2SO_4, H_2O, and then distill.
2. *Standards*: Stock: 2,000 ìmoles of palmitic acid/L = 0.5128 g/L of CHCb.

1,000 μmoles/L = 50 mL/lOO mL of CHCb

500 μmoles/L = 25 mLlI00 mL of CHCb

250 μmoles/L = 12.5 mL/lOO mL of CHCb

or

1,000 μmoles/L = 5 mL/10 mL of CHCb 500 ìmoles/L = 2.5 mL/10 mL of CHCb, 250 ìmoles/L = 1.25 mL/10 mL of CHCb

Procedure

1. To culture tube add 100 ul. of heptane (blank), standard, or plasma (run samples in duplicate).
2. Add 4 mL of chloroform:heptane:methanol solution and cap.
3. Shake for 5 min.
4. Add 1 mL of .035 N HCl and recap.
5. Shake for 3 min.
6. Centrifuge for 4 min at 1,000 x g
7. Aspirate and discard the upper aqueous methanol supernatant fluid.

   Note : do not disrupt the protein layer at the interface of the supernatant and lower phase.
8. Add 2 mL of cobalt reagent and recap.
9. Shake for 3 min.
10. Centrifuge at 1,000 ×g for 6 min.
11. Transfer 2 mL of upper supernatant liquid to clean culture tube as soon as possible

    Note: period should not exceed 30 min.
12. Add 1 mL of indicator and vortex.
13. Let stand for 30 min for color to develop. Color is stable for approximately 1 h.
14. Read at 435 nm in spectrophotometer.

Suggestion for analysis FFA

1. Use indicator within 2 d.
2. Use chloroform within 1 wk or redistill.
3. Make sure tubes are very clean.

4. Use a high pool and low pool in each run (high and low serum FFA).
5. Separation after cobalt addition may not occur - this requires a rerun.
6. 54 tubes at a time, which equates to 20 samples in duplicate.

Reference

Smith, S. W. 1975. A new salting out technique for calorimetric free fatty acids assays. Analytical Biochem. 67:531-539.

8.3.13. Transaminases

There are two enzymes involved in liver protein metabolism, e.g. Aspartate aminotransaminase (AST) or Glutamic oxaloacetic transminase (GOT; EC 2.6.1.1.) and Alanine aminotransaminase (ALT) or Glutamic pyruvic transaminase (GPT; EC 2.6.1.2).

The enzymes, GOT and GPT catalyses the following reactions.

GOT

Glutamic acid + oxaloacetic acid ? α-ketoglutaric acid + Aspartic acid.

GPT

Glutamic Acid+ Pyruvic acid ? α-ketoglutaric acid + Alanine

Principle

Colorimetric estimation of the enzyme activity is done in backward reaction. GPT activity is determined by measuring the amount of pyruvic acid released during incubation of the substrate with the enzyme. Pyruvic acid gives brown colour of hydrazone, formed by its reaction with 2,4-dinitrophenyl hydrazine (DNPH). In GOT, oxaloacetic acid is released during incubation of the substrate and the enzyme converts it spontaneously into pyruvic acid by decarboxylation. The pyruvic acid thus finally is estimated for measuring GOT acitivity (Reitman and Frankel, 1957).

Glutamic oxaloacetic transminase

Reagents

1. 0.1 mM phosphate buffer, pH 7.4.
2. GOT substrate: Dissolve 2.66 g DL-aspartic acid in a minimum volume of IN NaOH. Adjust pH to 7.4 and add 0.028 g

a-ketoglutaric acid. Dissolve it by adding little more 1 N NaOH. Adjust pH to 7.4 and make it to 100 ml with phosphate buffer. Store it frozen in small aliquots of 10 ml.

3. *GPT substrate*: Dissolve 1.8 g alanine in water and adjust pH to 7.4 by additing 1 N NaOH, Add 0.028 g a-ketoglutaric acid and dissolve it by adding a little more 1 N NaOH. Adjust pH to 7.4 and make it 100 ml with phosphate buffer. Store it frozen in small aliquots of 10 ml.
4. *4mM pyruvate standard*: Dissolve 11 mg sodium pyruvate in 25 ml phosphate buffer. Store frozen in 1 ml aliquots.
5. *ImM 2,4-dinitrophenylhydrazine (DNPH):* Dissolve 18.8 mg DNPH in 10 ml of concentrated hydrochloric acid and make upto 100 ml with distilled water. Store in brown bottle at room temperature.
6. *0.4N Sodium hydroxide*: Dissolve 16 g NaOH in distilled water and make upto 1 litre.

Procedure

1.
(a) *Test* : Take 0.1 ml sample and 0.5 ml substrate in a tube and incubate for 60 min at 39°C.
(b) *Control* : Take 0.1 ml sample and 0.5 ml substrate. No incubation is required.
(c) *Pyruvate standard:* Take 0.4 ml substrate, 0.1 ml pyruvate solution and 0.1 ml distilled water. No incubation is required.
(d) *Blank*: 0.5 ml substrate and 0.1 ml distilled water.

2. Add 0.5 ml DNPH immediately and leave the tubes for 20 min at room temperature.
3. Add 5 ml of O.4N sodium hydroxide.
4. Record optical density (OD) at 510 nm against reagent blank.

Calculation

Enzyme activity IS expressed as ìmol pyruvate produced per litre sample per min.

$$\frac{T-C}{S-B} \times \frac{0.4 \times 1}{60} \times \frac{1000}{0.1} = \frac{T-C}{S-B} \times 67\mu mol$$

Glutamic pyruvic transaminase

The procedure for GPT estimation IS same as for GOT with the following differences.

1. Use GPT substrate
2. Incubation time for (a) test is 30 min.
3. Enzyme activity.

Calculation

Enzyme activity IS expressed as ìmol pyruvate produced per litre sample per min.

$$\frac{T-C}{S-B} \times \frac{0.4 \times 1}{30} \times \frac{1000}{0.1} = \frac{T-C}{S-B} \times 133\mu mol$$

□□□

Chapter 9

Milk Composition, Analysis and Practical Considerations

The composition and functional properties of cow's milk are of considerable importance to the dairy farmer, manufacturer, and consumer. Broadly, there are three options available for altering the composition and/or functional properties of milk: cow nutrition and management, cow genetics, and dairy manufacturing technologies. Dairy cows derive fatty acids for milk fat synthesis from the diet/rumen microorganisms, from adipose tissues, and from *de novo* synthesis in the mammary gland. However, the relative contributions of these sources of fatty acids to milk fat production are highly dependent upon feed intake, diet composition, and stage of lactation. Feed intake, the amount of starch relative to fibre, the amount and composition of long chain fatty acids in the diet, and energy balance are particularly important. Significant differences in these factors exist between pasture-based dairy production systems and those based on total mixed ration, leading to differences in milk fat composition between the two. High intakes of starch are associated with higher levels of *de novo* synthesis offat in the mammary gland, resulting in milk fat with a higher concentration of saturated fatty acids. In contrast, higher intakes of polyunsaturated fatty acids from pasture and/or lipid supplements result in higher concentrations of unsaturated fatty acids, particularly oleate, trans-vaccenate, and conjugated linoleic acid (CtA) in milk fat. The concentration and composition of protein in milk are largely unresponsive to variation in nutrition and management. Exceptions to this are the effects of very low intakes of metabolisable energy (ME) and/or metabolisable protein (MP) on the concentration of total protein in milk, and the effeets of feeding with supplements/additives (organic Se, phytoadditives, prebiotics etc). However, low concentrations of protein in milk is sometimes, associated with seasonal variations in the nutritional value of herbage, may be a response to low intakes of both ME and MP. Stage of lactation is important in determining milk protein concentration, but has little influence on protein composition. The exception to this is in very late lactation where stage of lactation and low ME intake can interact to reduce the casein fraction and increase the whey fraction in milk and, consequently, reduce the yield of cheese per unit or milk. Research into targetted feeding strategies that make use of feed supplements including oil seeds, vegetable and fish oils, and organic Se supplements would increase the management options available to dairy farmers for the production of milks 'that di ffer in their composition. Given appropriate market signals, milk could be produced with lower concentrations of fat or higher levels of unsaturated fats, including CLA, and/or high concentrations of selenoproteins. This has the potential to allow the farmer to find a higher value market for milk and improve the competitiveness of the dairy manufacturer by enabling better matching of the supply of dairy products to the demands of the market.

9.1. Sampling of Milk for Analysis

A careful and accurate sampling of milk is of utmost importance in all analyses of milk. The whole milk from which a sample is to be taken should be uniform throughout in its composition. Before withdrawing samples for analytical determinations, bring the sample to a temperature of 26° to 28°C and mix thoroughly. However, milk should not be shaken violently under any circumstance as it may result in uneven distribution of fat. If lumps of cream do not completely disappear, warm the milk sample to about 400 C, mix thoroughly and then cool to26° to 28°C. All the samples shall be allowed to stand for three to four minutes after mixing to allow air bubbles to rise. The sample bottle then be inverted three or four times immediately prior to pipetting the milk for analysis. About 50 ml sample is taken in a numbered and stoppered glass tube daily during the trial period. Milk samples are stored in a deep freeze after adding 2-3 drops of potassium dichromate or mercuric chloride solution as a preservative for further analysis.

Random milk samples are collected from cows at different stages and duration of lactation and are usually analyzed for various milk constituents immediately after collection and in remote cases, may be kept in a refrigerator for part-time storage. The stored milk samples need to be thawed mixed well prior to analysis. The milk constituents (fat, total solids, protein, lactose and minerals) are analysed by following the protocol prescribed in AOAC (1990). Other parameters like pH, density (specific gravity), total milk yield and yield of different milk components, composition of milk fatty acids, lacto-albumin, etc have also significance to define qualitative and quantitative milk production.

9.2. Analysis of Milk Constituents

9.2.1. pH

The pH of milk can be measured by a hand-held pH meter at the milk collection site or in table-top pH meter after bringing the milk to laboratory.

9.2.2. Density (Specific gravity)

Lactometer or automatic milk analyser is used to measure specific gravity of milk.

9.2.3. Determination of ash content in milk

The ash is made up of oxides and chlorides of the mineral elements. Milk ash is known to contain Potassium, Sodium, Calcium, Magnesium, Chlorine, Phosphorus and Sulphur in relatively large amounts. Small amounts of Iron, Copper, Zinc, Aluminium, Manganese and Iodine and traces of Silicon, Vanadium, Boron, Titanium, Rubidium, Lithium and Strontium have been also reported. The ash is white to gray matter obtained by evaporating the milk and then by igniting the dry matter. It is most constant constituent of milk and its content ranges between 0.7 to 1.0 per cent.

Apparatus

Crucible (Platinum or silica), Pipette 10 ml capacity, Balance, Water bath, Muffle furnace, Desiccator

Procedure

1. Take the crucible, which has been dried, cooled in a desiccator and weighed previously.
2. Put 10 ml of prepared sample and weigh.
3. Evaporate the milk to dryness on a water bath.
4. Ignite the dried residue in a muffle furnace at a temperature not more than 550°C until the ash is free from carbon.
5. Cool in a desiccator and weigh.

   Note: Ash may be dissolved in diluted hydrochloric acid and volume is made to 100 ml in a graduated volumetric flask for its further analysis.

Calculation

Percentage of ash by weight= (100 w)/W

Where, w = Weight of ash (g) W= Weight of milk (g)

9.2.4. Determination of total solids in milk

The entire residue left after complete evaporation of water from milk is known as dry matter or total solids (TS). The term total solids of milk include fat, proteins, lactose and mineral matter. The amount of total solids in milk varies within the considerable limits and is greatly influenced by fat content of milk, higher the milk fat content greater the amounts of total solids. Total solids minus milk fat, constitute, as the name implies, the solidsnot-fat (SNF). Generally, for each per cent increase in milk fat content there is corresponding increase of about 0.4 per cent in SNF. The total solids of milk is usually determined by the gravimetric method, it may however be calculated rapidly by a formula or more rapidly by using Richmond's milk scale.

By Evaporation (Gravimetric Method)

Apparatus

Crucible (platinum, silica, nickel, aluminium, porcelain etc.) not less than 5 cm in diameter, Pipette of 10 ml capacity, Balance, Water bath, Hot air oven, Desiccator.

Procedure

1. Take the crucible, which has been dried, cooled in a desiccator and weighed previously.

2. Pipette into the dish 5 ml of milk and weigh quickly so that constant evaporation could be avoided.
3. Place the crucible on boiling water bath.
4. After at least 30 min. remove the crucible from the water bath and place it in the hot air oven at 900 to 100°C until the constant weight.
5. Cool in a desiccator and weigh.

Note : Crucibles with total solids can be used for further analysis of ash and mineral matter.

Calculation

Percentage of TS by weight = (100 w)/W

Where, w= weight of the residue after drying (g) W= weight of the milk (g).

By Calculations

1. Fleischmann's Formula: Percentage TS = 1.2F+ 2.665*(100 S - 100)/S; SNF = TS - F
2. Richmond's Formula: Percentage TS= (L/4) + 1.2F+ 0.14; SNF = (L/4) + 0.2F+ 0.14
   Where, F = percentage fat in milk S = Specific gravity of milk, L = Lactometer reading at 15.5°C

Alternate method (Dry matter)

1. In a clean and dry beaker of 50 ml capacity, put about 5 cm thick layer of absorbent cotton and dried in hot air oven at 65 °C for 48h, cooled and weighed (W1).
2. Exactly, 10-20 ml of milk sample is pipetted in to the beaker to be absorbed by the cotton pad, weight recorded (W2) and is put in the hot-air oven again for drying at 65 °C for 48h or more until a constant dry weight. Prior to weighing, the beaker is cooled to room temperature in a desiccator and weighed again (W3).
3. The total solid (dry matter) content (%) is determined as 100 × (W3-W1)/(W2-W1).

N.B. In place of cotton, washed silica may also be taken for the purpose.

9.2.5. Determination of fat in milk

Milk fat measurement is a common task in the dairy industry, because milk fat content is one factor that determines milk price and is necessary to know for casein/fat ratio normalization. It is also important for the dairyman to know milk fat content exactly: discrepancies in the results of milk fat tests (usually performed in the dairy industrial plant) have economic relevance. A low or high milk fat content also speaks for diet type, nutrition and health status of a milch animal.

By Gerber's method

The Gerber test for the determination of percentage of fat in milk was devised by Dr. N. Gerber of Zurich, Switzerland. The milk fat is separated from milk by strong acidifying and rotating the milk in a specially designed Gerber's acid-butyrometer in the Gerber's centrifugal machine. The method is simple, quick and capable of high degree of accuracy if carried out carefully. In India this test is most commonly used for estimation of fat content in milk.

Principle

In this method the casein of milk is dissolved with sulphuric acid (10: 1) and separation of fat is facilitated by the addition of amyl alcohol. Addition of amyl alcohol prevents the partial charring of the fat and sugar by sulphuric acid, keeps the fat clear and renders the separation of fat easier.

Apparatus

Gerber's butyrometer, Pipette of 11 ml capacity for milk, Automatic acid measuring apparatus (10 ml capacity), Pipette 1 rnl capacity for amyl alcohol, Wodden or metal stand for butyrometer, Water bath., Gerber's centrifuge (1100 rpm).

Reagents

1. Sulphuric acid (10: 1) Specific gravity 1.820 to 1.825.
2. Amyl alcohol specific gravity 0.815 to 0.818.

Procedure

1. Mix the milk sample (temperature about 20°C) thoroughly, taking care to minimize incorporation of air. Allow the sample to stand

for a few minutes to discharge any air bubbles. Mix gently again before pipetting.

2. Place the clean and dry butyrometer on stand with open ends upwards.
3. Pipette sequentially into the butyrometer: 10 ml of 90% sulphuric acid, 10.75 ml milk (must not be mixed with the acid, but gently layered over it down the inside of the butyrometer to avoid charring), 1 ml amyl alcohol, some drops of distilled water
4. Clean the neck of the butyrometer with a tissue or dry cloth and close the butyrometer with caoutchouc stopper
5. Shake the butyrometer with milk and reagents until the milk was dissolved and turned upside-down for 5 or more times. When mixing, hold the butyrometer stopper firmly to ensure that it cannot slip.
6. Spin in the Gerber centrifuge at 1100 rpm for 15 minutes.
7. The fat column was adjusted by using the stopper, so that it will be in the graduated part of the butyrometer. The fat percentage was read directly.
8. Remove and place the tube in a water bath at 68°C for 2-3 min.
9. Take out the butyrometer from the water bath and adjust the stopper to bring the lower end of fat column against a unit graduation and read from there to the lowest part of the upper meniscus.

9.2.6. Determination of total nitrogen

Principle

Organic matter when digested with concentrated sulphuric acid in the presence of catalyst is converted to ammonium sulphate. Ammonia liberated by making the solution alkaline is distilled into known volume of standard acid is then back titrated.

Apparatus

Kjeldahl digestion flask, Digestion bench, Volumetric flask, Auto Kjeltec

Reagents

1. Concentrated sulphuric acid.
2. Mercuric oxide
3. Potassium sulphate.
4. Sodium hydroxide.
5. 0.1 N Standard sulphuric acid solution.
6. Boric acid solution with indicator

Procedure

1. Place about 5 g of milk in a 500ml Kjeldahl digestion flask.
2. Add 265 ml of concentrated sulphuric acid.
3. Add 0.7 g of mercuric oxide and 15 g of powered potassium sulphate.
4. The digestion is started with a temperature below boiling point and heat is gradually increased.
5. Finally, use full flame and complete the digestion.
6. Determine the nitrogen content according to Kjeldahl method as discussed earlier.

Determination of crude protein

Crude protein percentage is obtained by multiplying total nitrogen content with a factor 6.38.

Note: The factor 6.38 is used to calculate the protein content from the total nitrogen, since casein, the principle protein of milk, has a nitrogen content of 15.7 per cent.

9.2.7. Determination of true protein

Reagents

1. Trichloroacetic acid solution 15 percent by volume.

Procedure

1. Pipette 10 ml of milk into 50 ml tared graduated flask and weigh accurately.
2. Dilute to the mark with trichloroacetic acid solution and mix immediately.

3. Filter the precipitate with the help of filter paper (Whatman no. 40).
4. Wash the precipitate with the trichloroacetic acid solution.
5. Determine the nitrogen content in the washed precipitate and paper.

Calculation

Obtain the percentage of true protein by multiplying the percentage of nitrogen by the factor 6.38.

9.2.8. Determination of non-protein nitrogen

Reagents

1. Trichloroacetic acid solution 15 percent by volume.

Procedure

1. Pipette 10 **ml** of into 50 ml tared graduated flask and weigh accurately.
2. Dilute to the mark with trichloroacetic acid solution and mix immediately.
3. Filter the precipitate with the help of filter paper (Whatman no. 40).
4. Wash the precipitate with the trichloroacetic acid solution.
5. Determine the nitrogen content in 20 ml of filtrate.

Calculation

Calculate the nitrogen content considering of weight of sample and filtrate obtained.

9.2.9. Determination of casein

Method I (Hont's method)

Apparatus

Pipette, Erlenmeyer flask 200 ml cap., Filter paper (Whatman no. 40), Burette

Reagents

1. Acetic acid 10 per cent by volume
2. 0.1 N Potassium hydroxide
3. 0.1 N Hydrochloric acid
4. Phenolphthalein indicator

Procedure

1. Take 10.5 ml of milk into a 200 ml Erlenmeyer flask.
2. Add 75 ml of distilled water and 1-1.5 ml of 10 per cent acetic acid.
3. Mix the contents with gentle rotation and set aside for 10 minutes.
4. Filter the precipitated casein with filter paper (Whatman no. 40).
5. Wash out the absorbed and loosely bound acetic acid by means of cold water.
6. Continue washing of casein on filter paper and that adhering to the flask until the volume of washed water reaches to at least 250 ml.
7. Transfer back the precipitate and the filter paper to the original Erlenmever flask.
8. Add 75-80 ml of C02 free distilled water and 10 ml of 0.1 N KOH and a few drops of phenolphthalein indicator.
9. Stopper the flask and shake it vigorously until the casein has been brought into the solution.
10. Rinse the stopper with free water and titrate the alkaline casein solution at once with 0.1 N HCl till the red colour disappears.
11. Run a blank test parallel to the entire determination. This check titration should be added to the volume of acid used in titration.

Calculation

Subtract the corrected acid reading from the 10 ml of alkali used. The difference is percentage of casein in the milk.

Method II

Reagents

1. Acetic acid 10 per cent by volume.
2. 1 N Sodium acetate solution.
3. Filter paper (Whatman no. 40)

Procedure

1. Take 5 ml of milk and weigh accurately into a 100 ml beaker.
2. Add 50 ml of distilled water and 0.5 ml of 10 per cent acetic acid.

3. Mix the contents with gentle rotation and set aside for 10 minutes.
4. Add 0.5 ml of sodium acetate and mix again.
5. Allow the contents of beaker to cool and decant into a pleated filter paper.
6. Wash the precipitate with water three times, and by decantation, transfer it to the paper and wash it twice on the paper.
7. Determine the nitrogen content in the washed precipitate and paper using 25 ml of 0.1 N sulphuric acid solution and a paper in addition to 5 ml of water for the blank determination.

Calculation

Calculate the percentage of nitrogen and multiply by 6.38 to obtain the percentage of casein.

9.2.10. Determination of albumin

Reagents

1. Acetic acid 10 per cent by volume.
2. 1 N Sodium acetate solution.
3. Filter paper (Whatman no. 40)
4. Sodium hydroxide solution 10 per cent w/v.

Procedure

1. Take 5 ml of milk and weigh accurately into a 100 ml beaker.
2. Add 50 ml of distilled water and 0.5 ml of 10 per cent acetic acid.
3. Mix the contents with gentle rotation and set aside for 10 minutes.
4. Add 0.5 ml of sodium acetate and mix again.
5. Allow the contents of beaker to cool and decant into a pleated filter paper.
6. Wash the precipitate with water three times, and by decantation, transfer it to the paper and wash it twice on the paper.
7. Neutralize the filtrate with sodium hydroxide solution.
8. Add 0.3 ml of dilute acetic acid and heat on a steam bath until the albumin is completely precipitated.
9. Collect the precipitate on a acid washed filter paper and wash with cold water.
10. Determine the nitrogen in the precipitate.

Calculation

Multiply the percentage of nitrogen by 6.38 to obtain the percentage of albumin.

9.2.11. Determination of lactose

Principle

To determine the lactose content, protein free filtrate of milk is heated with alkaline copper solution using a special tube (Folin-Wu) to prevent reoxidation. The cuprous oxide formed is treated with phosphomolybdic acid solution and a blue colour is being obtained which is compared with that of standard.

Apparatus

Folin-Wu tube, Volumetric flask,

Reagents

1. Sodium tungstate 10 per cent.
2. 0.66 N Sulphuric acid.
3. Folin-Wu alkaline copper solution: Dissolve 40 g pure anhydrous sodium carbonate in about 400 ml of water and transfer to a litre flask. Add 7.5 g of tartaric acid and when the latter is dissolved add 4.5 g of crystallized copper sulphate. Mix and make up to a volume of 1 litre.
4. Phosphomolybdic acid solution: To 35 g of molybdic acid and 5 g of sodium tungstate add 200 ml of 10 per cent sodium hydroxide and 200 ml of water. Boil vigorously for 20 min. so as to remove nearly the whole of the ammonia present in the molybdic acid. Cool, dilute to about 350 ml, then add 125 ml of cone. (85%) phosphoric acid and dilute to 500 ml.

Procedure

1. Take 1 ml milk into a 100 ml volumetric flask.
2. Add 2 ml of 10 percent sodium tungstate.
3. Add 2 ml of 0.66 N sulphuric acid gradually, mix well and allow to stand for 5 minutes, thereafter, dilute to the mark and filter.
4. Introduce 1 ml filtrate and 1 ml water in a Folin-Wu tube. Take 2 ml of working standard lactose solution in another tube.

5. Add 2 ml of alkaline copper solution to each tube and heat in boiling water for eight minutes.
6. Cool the contents and add 4 ml of phosphomolybdic acid solution to each tube.
7. Add diluted phosphomolybdic acid solution (1:4) to the 25 ml mark after one minute.
8. Mix the contents well and read the *a.D.* in Spectrophotometer at 420nm against blank.

Calculation

$$\text{Lactose (\%)} = \frac{\text{OD of unknown} \times 0.6 \times 100}{\text{OD of known} \times 0.01 \times 1000}$$

Alternate method

The lactose in milk or whey is determined by Teles' method (Teles *et al.,* 1978).

Reagents

1. 4.5% barium hydroxide
2. 5% zinc sulphate
3. 1% phenol
4. 5% sodium hydroxide
5. 1% picric acid
6. 1% sodium bisulphate

   N.B. All the solutions are prepared in distilled water w/v
7. *Teles' reagent (working solution):* mix in the following order: 1 vol phenol, 2 vol sodium hydroxide, 2 vol picric acid, 1 vol sodium bisulfite solutions, mix well after each addition and keep in dark container)

Procedure

1. Add 0.2 ml 5% zinc sulfate to 2.5 ml aliquot of the diluted (2% v/v with distilled water) milk or whey sample in a centrifuge tube and then add 0.2 ml 4.5% barium hydroxide
2. Centrifuge for 1 min at 1000 rpm
3. Transfer 1.0 ml supernatant in to a Folin blood sugar tube or any glass tube and add 2.5 ml Teles' reagent

4. Keep in a boiling water bath for 6 min (for other glass tube, put a glass bulb to avoid escape of vapour maintaing the volume of reaction mixture)
5. Coll immediately in cold tap water and make the volume to 12.5 ml with distilled water and mix well
6. Read absorbance at 520 nm against a similarly treated reagent blank having 2.5 ml water in place of diluted sample
7. Compare with a standard solution of lactose (1 mg/ml; 0.1052 g lactose in 100 ml distilled water) subjected to similar treatment as that for sample and blank

9.2.12. Milk analysis by automatic analyzer 'Laktan 1-4" model 220'

The autoanalyzer *'Laktan 1-4" model 220* is used to determine the percentage of fat, protein, solids-non-fat (SNF), added water, temperature and density in a sample of fresh whole, preserved, pasteurized, normalized, powder milk, skimmed milk and a long storage milk. It takes 130 seconds to determine these parameters and doesn't require any chemical reagents. The value of this device is significant, because one traditional method of protein determination takes 6 hours and additional materials. It saves a lot of time and money and can be used during the milk reception.

- The machine is put on for 10 min prior to calibration.
- About 20 ml milk is poured in to the container and set for reading by putting the knob.
- The display will give the details of above parameter.
- For accurate reading, the knob should be washed and the analyzer is calibrated for each and every different sample of milk.

References

IB.I. 1961. Indian standard Methods of Test Dairy Industry. Part II. Chemical Analysis of milk. IS: 1479 (Part II). Indian standard Institution, New Delhi.

Teles, F.F.F., Young, C.K. and Stull, J.W., 1978. A method for rapid determination of lactose. J. Dairy Sci. 61, pp. 506–508.

□□□

Chapter 10

Statistical Methods

Mean

Making tables and graphs are always the first step in attempting to summarise a set of data and to look at its structure. One of the first items of interest is a measure of a central or representative overall value. The most commonly used measure is the arithmetic mean (usually referred to simply as the mean). The mean is obtained by dividing the sum of all the observations by the number of observations (denoted as $\mu = Sx_i/n$). However, the mean can be somewhat misleading if the distribution is skewed, i.e. if there is a high proportion of large values or a high proportion of small values. The problem is that a few extreme values unduly influence the value of the mean. In these situations the median may be a very useful alternative. The median is the value lying in the middle, such that one half of the values are above and one half below. Suppose the pre-weaning body weights of lambs are put in increasing order say 8.4, 9.6, 12.1, 13.2, 14.1, 16.3, 17.2, 17.7 kg; the median value is 13.65 kg. If there is odd number of values, the median is the middle value. In both cases, the numbers must be arranged in ascending order (from smallest to largest) before calculating the median. These two values (for mean 13.575 and median 13.65) are in most cases similar.

Variance, Standard Deviation and Standard Error

Biological processes yield variable data. The most common measures of variability are the variance and standard deviation. The degree to which numerical data tend to spread about or differ from a mean and is calculated as $? = \Sigma\sqrt{\{(\mu - x_i)^2/n\}}$. The standard error of mean is $?/\sqrt{n}$. The variance of a group of observations is calculated as the sum of squares of deviations from the mean divided by one fewer than the number of observations ($?_{n-1} = \Sigma\sqrt{\{(\mu - x_i)^2/(n-1)\}}$. An alternative formula, which is easier to use with a large number of observations, is to calculate the sum of each value squared, subtract the total squared divided by the number of observations, and divide the answer by the number of observations less one. There are occasions when we wish to compare the relative variation between two variables having different means, e.g. birth weights of lambs and adult weights of ewes. Since the adult weights would be expected to have a larger standard deviation because of the larger values involved, a comparison of the two standard deviations may not be very helpful. We might also be interested in comparing the relative variation between different characteristics or traits of the same animals, e.g. weaning weights and PCV of lambs. For such comparisons the coefficient of variation (CV) is calculated as the standard deviation expressed as a percentage of the mean.

Suppose the experiment were repeated several times. Each time we would calculate a mean, say for breed 1, diet 1, each mean would be different. Thus there would be variability among means just as there is variability among individual observations in each sample. However, since each mean is averaged over eight observations, the variation among means would be smaller than the variation among individual observations. The variance of a mean can be calculated as the average of the variances calculated each time the experiment is conducted divided by the number of observations used for calculating each mean. The standard deviation of the mean is known as the standard error (i.e. $?/\sqrt{n}$.). The standard error or variance of a mean decreases as the number of observations increases.

Confidence Interval

The standard error can be used to develop what is known as a confidence interval. A confidence interval is a range between upper and lower limits, which is expected to include at a given level of probability the true (or population) mean value. This is the value for which the sample in the experiment provides an unbiased estimate.

This is the interval in which the true mean should lie with, say 95%, or 19 times in 20, chance of being correct. Similarly, the 99% confidence interval gives the range within which we expect the true mean to lie with a 99%, or 99 times in 100, chance of being correct. The approximate 95% confidence interval can be calculated as the sample mean plus or minus twice the standard error. Similarly, the 99% confidence interval is the sample mean plus or minus approximately 2.6 times the standard error.

Thus the approximate 95% confidence interval for the mean 6 mo body weight of lambs of say breed 1 fed diet 1 is 30.58 ± t-value at 5% (say 2.0) × SE (say 1.11) = (28.36 to 32.80 kg).

Similarly, the approximate 99% confidence interval is 30.58 ± t-value at 1% (say 2.6 × 1.11 = (27.69 to 33.47 kg).

As one might expect, the 99% confidence interval is wider than the 95% one. Another way of thinking about this is to say that if the experiment were repeated 100 times then we would expect that on 99 occasions the sample mean would fall within the range 27.69 to 33.47 kg, and on the other occasion it would fall outside. It should be emphasised that the above formula is only approximate; instead of 2 or 2.6 we should strictly use a t-value with the appropriate degrees of freedom.

Binomial and Poison Variable

The variables are sometimes described as discrete variables, *i.e.* they can only take on discrete values. Variables such as body weight are known as continuous variables, *i.e.* they can take on any value within a reasonable range to a given degree of accuracy. Discrete variables such as mortality and disease often belong to distributions that are not normal, but binomial or Poison distributions. Discrete variables taking on values 0 or 1 are also sometimes called binary variables. For example, suppose that 4 lambs in a sample of 20 die before weaning. We can record a death as 1, a lamb that survived as

O. The proportion of lambs that died is p = 4/20 = 0.2. This is simply the mean of the Os and Is, i.e. (1 + 1 + 1 + 1 +0 .... +0)/20 = 4/20 = 0.2. Mortality is typical of a variable that is associated with a probability of occurrence, p. Such variables typically belong to a binomial distribution.

We have seen that the mean = p, and that this can be calculated in an analogous way to that of a normally distributed variable. The binomial distribution, however, has a special expression for the variance, namely p(1- p). Thus if p = 0.2, then variance = 0.2 × 0.8 = 0.16.

The standard deviation and standard error are similarly calculated, i.e. standard deviation = v0.16 = 0.4; standard error = v(0.16/20) = 0.089.

The distribution of certain binary data is often closer to that of a Poisson distribution, a distribution associated with rare events. Incidence of trypansomosis, for example, can often be considered as being associated with a Poisson distribution, particularly if the prevalence is very low (e.g.<lO%). The mean is the same as that for the binomial distribution, namely p. The variance, however, is different from that of the binomial distribution and is also p. Thus, if the prevalence of trypanosome infections is 0.06 in a sample of 100 cattle, (i.e. 61100 are detected parasitaemic), then the variance is also 0.06 and standard deviation = v*0.16* = 0.4; standard error = v(0.06/100) = 0.024. In practice when p is small (<0.1), it does not matter whether one assumes a binomial or Poisson distribution.

Analysis of Variance

The purpose of analysis of variance is to estimate the overall significance of differences among each set of parameter means. It separates variations due to breed and diet from each other, and compares the magnitudes of these different sources of variation with the variation which is left over among residual terms. In developing the idea of analysis of variance it will be simplest to start with a model based on different groups of lambs categorized by breed and diet. It is important to emphasize the need for writing down the statistical model to be fitted to a set of data before embarking on a statistical analysis. It helps the researcher decide the form the analysis of variance should take. In this expression, an observation can be defined to have an overall mean, a term for breed, a term for diet and a residual. These equations can be written more generally as:

$$Y_{ijk} = \mu + b_i + d_j + e_{ijk} \; (i=1-n; j = 1,-n, k= 1,--n)$$

In using analysis of variance, two main assumptions are used: firstly, that the eijks have the same distribution for each parameter or factor (i.e. are not more variable for one than another) and, secondly, they are distributed normally. Thus, it is not strictly valid to use analysis of variance for discrete variables that are associated with binomial or Poisson distributions.

One-way analysis of variance

Say, four groups of eight lambs each categorized by breed and diet in a data set. The model "Yij =μ,+g1+eij" is an example of one-way analysis of variance and such an experiment is often referred to as a completely randomized design.

The degrees of freedom (df) for the variation among groups is calculated as one fewer than the number of groups (i.e. 4-1). The total degrees of freedom = 32 (number of lambs)-1. The residual degrees of freedom is calculated by subtracting the among groups df from the total df.

The following calculations calculate the sums of squares (SS) and mean squares (MS).

Total. This is the same calculation as used previously for the variance, i.e. the sum of squares of individual observations minus the square of the total divided by the number of observations. Mathematically we can write this as:

$$\Sigma Yij^2 - (\Sigma Yij)^2/32$$

Thus, if we apply this formula to the weight gains of the 32 lambs, we obtain total SS. The formula for the among group sums of squares uses the totals for each group divided by the number of lambs in each group. It is:

$$\Sigma Yi.^2/8 - (\Sigma Yij)^2/32$$

The capital letter Y, together with the dot, means the sum of the individual yijs summed over the subscript that the dot is replacing, in this case, j. Thus:

$$Yi. = Y_{11} + Y_{12} + ... + Y_{18}$$

The among group mean square equals the among group sum of squares divided by the among group degrees of freedom. The residual sum of squares is calculated by subtracting the among group sum of squares from the total sum of squares. The residual MS estimates the average variation among individuals within groups and is an estimate of the average within group variance of weight gains. The among group MS represents the additional variation brought about by the differences in mean weight gains among the four groups. The magnitude of this additional variation can be assessed by dividing the among group MS by the residual MS to give what is known as the F-value. We use this value to compare with values in F-tables with 3 and 28 degrees offreedom. Ifwe look at the column headed 3 df at the top of each of the F-tables and then come down this column until we reach 28 df on the left hand side, we find the values 2.95, 4.57, and 7.19 for 5%, 1 % and 0.1 % levels ofprobability, respectively. If our observed F-value exceeds any of these values, then we say it is statistically significant.

Factorial analysis of variance

This experimental design consists of two or more factors, each with discrete possible values or "levels", and whose experimental units take on all possible combinations of these levels across all such factors. The groups of lambs in the experiment have been assigned a particular structure. In other words the 4 groups are composed of combinations of 2 diets and 2 breeds, namely:

	Supplementation	
Breed	**Yes**	**No**
1	1	2

This design, sometimes described as a 2x2 factorial design, is a very useful way of getting as much information from an experiment as possible. Thus we can look at the effects of dietary supplementation on weight gain of two breeds at the same time. What is more, we can look at the way that dietary supplementation and breed may interact. If there is no interaction then we mean that any effect of dietary supplementation is the same for each breed. If there is an interaction then the effect of supplementation is different for different breeds.

We can rewrite our statistical model as follows:

$$y_{ijk} = \mu + b_i + d_j + bd(ij) + e_{ijk} \ (i=1,-n;\ j = 1,-n,\ k= 1,--n)$$

where (bd)ij is a notation used to signify the interaction of breed and diet.

If the analysis shows there to be no significant interaction then we could express the model as:

$y_{ijk} = \mu + b_i + d_j + e_{ijk}$

In other words, the model simply adds to breed the same estimate for diet irrespective of breed.

Sometimes we refer to such effects as being additive.

To estimate sums of squares for breed and diet we split the sums of squares we have already obtained for groups, namely:

$\Sigma Yij.^2/8 - (SYijk)^2/32$ into a component for breed:

$\Sigma Yi..^2/16 - (SYijk)^2/32$

and a component for diet:

$\Sigma Y.j.^2/16 - (SYijk)^2/32$

This analysis demonstrates a highly significant difference between breeds in average weight gain, a highly significant effect of supplementation with cottonseed cake and bran meal and an interaction approaching statistical significance. But how precisely were these effects manifested? We need to look at the mean values and calculate their standard errors to determine the precision with which they are estimated. We can then apply earlier knowledge to calculate confidence limits for each mean. Standard errors (SE) can be calculated as the square root of the residual variance divided by the number of observations.

t-test

Instead of the F-test, which considers the variation among any number of levels of a factor, we use the t-test which looks at the variation between just two levels.

The formula for the t-test is:

t = (difference between two means)/(SE of difference between two means)

The SE of the difference between two means (SED) is calculated as the square root of the variance multiplied by (1/n1 + 1/n2) where n1 and n2 are the numbers of observations respectively making up the two means to be compared. Thus, SED = [variance × (1/n1 + 1/n2)]. When n1 = n2, SED = v(2× variance of mean) = v × SE. The observed value is compared with t-values in the t-table using the number of degrees of freedom for the residual variance at say 5%,1 % and 0.1 % levels of significance and considered significant at that levels if the value is greater than the t-value in the table.

Least significant difference

It is sometimes more convenient to calculate the least significant difference (LSD) needed for the difference between two means to be significant. This is calculated as:

LSD = t X SED

Thus, we can then compare differences between the respective means with these values to determine their significance.

Randomised block design

One aim of good experimental design is to ensure that the residual mean square is as small as possible. One way to do this is to ensure that animals are as uniform as possible before the experiment starts. Another is to ensure that animals with similar attributes are assigned equally to the different factors being compared. The most common type of design is a randomised block for which animals might be blocked or grouped, for example, on the basis of body weight, age, lactation, pregnancy status or herd. An experiment might also be replicated in time. Each replicate then constitutes a block.

Because of the wide range of weaning weights among lambs in an experiment it might be sensible to group the lambs on the basis of weaning weight. Lambs in each breed were sorted into weaning weight order and formed into four blocks of four with the lightest four in the first and heaviest four in the fourth block. Lambs were then randomly assigned to the two levels of diet within each block. In this, the model formula becomes (with r for block):

$y_{ijk} = \mu + b_i + rik + d_j + bd(ij) + e_{ijkm}$ (i=1,—n; j = 1,—n, k= 1,— —n, m=1,—n))

Continuing with the example stated earlier, say lambs belong to two breeds, the formula for calculating the among block within breed sum of squares is the sum of

$\Sigma Y1jk1^2/4$ - $(SY1jkm)^2/16$ calculated for lambs of breed 1

$\Sigma Y2jk1^2/4$ - $(SY2jkm)^2/16$ calculated for lambs of breed 2

The new residual sum of squares is then calculated by subtracting this sum of squares from the residual sum of squares previously calculated.

Heterogeneity of variance

The variances for weight gains among individuals can be calculated separately for each breed-diet group and the residual variance used in the analysis of variance is the average of these. The analysis of variance we have undertaken assumes that each variance calculated for each group estimates the same 'population' variance. This may not be true. The analysis of variance technique handles data for which assumptions, such as homogeneity of variance of symmetric, normal distributions, do not strictly apply. But what alternative methods are at our disposal? Faecal egg count (FEC), for instance, has a very skewed, non-normal distribution and an analysis of variance undertaken on these data may not be justified.

Transformations: Firstly we could transform FEC to a different scale by using a calculation such as log (y), log (k−y) or vy. Each of these calculations reduces the skewness of the data and makes the residual variance less dependent on the mean. The logarithm transformation is stronger than the square root one. The log (k-y) transformation, where k is a constant >0, is used instead oflog (y) when individual values of y occur which equal O. The disadvantage of this method is that, although it may produce a more valid statistical analysis, it is often not so easy to present the results. This is because the means and standard errors calculated in the statistical analysis are now on a different scale. Sometimes a transformation is not appropriate and an alternative method would be to split the data into separate parts and to do a separate analysis of variance on each part. It is conventional, however, to analyze FEC on the log scale. An arcsin ($\sin^{-1}$vy) or square root transformation is more appropriate for disease prevalence data, the square root for when the prevalence is low. However, there are now new methods using what is known as logistic regression or log linear models which are more suitable for such data.

Results can be expressed on the original scale by calculating the geometric mean (antilog of the transformed mean) and the antilog of the 95% confidence limits for the transformed mean. Take, for example, the Dorper lambs (breed 1) supplemented with cottonseed cake and bran meal. The mean and standard error for FEC calculated on the log scale ($\log_e$ y) are 7.26 and 0.36, respectively. The 95% confidence range for this mean is 7.26 ± t X 0.36 where t, with 7 degrees of freedom = 2.36. Thus, 95% confidence interval for mean = (6.41, 8.11). If we transform these values back to the original scale using the antilog (or e^x) transformation, then mean = 1422, range = (608 to 3327).

Thus, we can say that the geometric mean is 1422 eggs per gram (epg) and that the 95% confidence interval within which the true 'population' mean lies is between 608 and 3327 epg. Three important points need to be highlighted:

- it is not permitted to calculate an antilog for the standard error to produce a standard error on the original scale
- the upper and lower confidence interval values are not equidistant from the mean
- if $\log_e$ (50 + y) is used for the transformation, then 50 must be taken off the antilogs after transformation back to the original scale in order to calculate the correct geometric mean and range.

Another variable that is often transformed to logarithms is antibody titre. Serum is usually diluted in a geometric series, i.e. with a constant ratio between successive dilutions, e.g. 1/2, 1/4, 1/8, 1/16 etc. Thus, frequency distributions of titres tend to be lognormal and so the logarithmic transformation is to be desired.

Regression analysis and analysis of covariance

Regression analysis is a form of analysis of variance. The only difference is that the statistical model is written in a form that is based on the relationship between a y variable and one or more other continuous variables, say x, z .... , often referred to as independent variables, rather than with discrete sets of parameter or factor levels.

Thus Yi = a + bx; + ei fits a straight line between the Yi and Xi values with slope b, known as the regression coefficient, and crossing the y-axis at the point a, known as the intercept. We shall not consider methods of calculation here, but they result in a similar analysis of variance table to those derived previously. For example, consider the relationship between PCV and $\log_e$ (50 + FEC) and ignore the fact that lambs were from different breeds and fed different diets. Statistical analysis of these data may lead to a different set of analysis of variance.

Analysis of covariance

Analysis of covariance is a simple extension of analysis of variance. Independent variables not incorporated in the design of the experiment are included as covariates. Thus, in the study of the effect of supplementation on the weight gains of the two breeds of sheep, weaning weight was taken into account in the definition of blocks. We can include sex in the model as a covariate with values 1 or 2 for males and females, respectively. Covariates may also be continuous variables. The full model appears as follows:

$$Yijkm = \mu + bi + rik + d_j + (bd)ij + ßxijkm + eijkm$$

where ß is the regression coefficient of weight gain (yijkm) on sex (xijkm).

The analysis adjusts, or corrects, individual parameter level means for the differences in numbers of males and females making up the means. The regression on sex, whilst not significant, has nevertheless reduced the residual mean square. Interestingly, when adjusted for sex, the F-value for blocks is now also greater than 1. Note that the total sum of squares is no longer the sum of the individual squares in the analysis of variance table. This is because each sum of squares is now corrected for differences between sexes. When there is a choice to use an independent variable as a covariate or to define it as a blocking factor in a designed experiment, the second option is usually preferred, provided adequate numbers of degrees of freedom are retained for the residual term. If there had been equal numbers of females and males for each breed in Table 8.1, then it would have been possible to block lambs on the basis of body weight for each sex.

Analysis of discrete data

Sometimes response variables measured in an experiment may be discrete, not continuous. Such variables, often referred to as binary variables, can take two values, 0 or 1. One example is mortality; others are proportions of animals conceiving, proportions lambing, or proportions with a given disease etc. This module briefly mentions the simplest method of handling such data.

Chi-square (c2) test

For simple experiments one can group the data into a two-way table and perform a c2 test. Thus, for example, if the lambs used in this experiment were drawn from a group of 55 Dorper and 42 Red

Maasai lambs and 14 and 4, respectively, died before weaning, then we can compare differences in mortality by forming a table, often known as a contingency table, as follows:

	Alive at weaning		
Breed	**No**	**Yes**	**Total**
Dorper	14	41	55
Red Maasai	4	38	42
Total	18	79	97

The above values are the ones observed. The next step is to calculate the values that would have occurred had the mortality rates in the two breeds been the same. These are known as expected values. To do this calculation, expected values for each of the four cells in the body of table are obtained by multiplying together the two totals in the corresponding row and column and dividing by the grand total. Thus, expected values are

55× 18/97	55× 79/97
42× 18/97	42× 79/97
10.2	44.8
7.8	34.2

The formula for the c2 test is:

$\Sigma(oi - ei)^2/ei$

where oi = observed value and ei = expected value (i = 1, 2, 3, 4).

Thus,

$$c2 = (14 - 10.2)^2/10.2\ (41 - 44.8)^2/44.8 + (4 - 7.8)^2/7.8 + (38 - 34.2)^2/34.2$$
$$= 1.416 + 0.322 + 1.851 + 0.422$$
$$= 4.01$$

The number of degrees of freedom = (2 - 1) × (2 - 1) = 1

The 5% and 1% values for c2 with 1 degree of freedom are 3.84 and 6.63, respectively. Thus we can say that the mortality rate before weaning in the Dorpers of 14/55 (25%) was significantly higher than that of 4/42 (10%) in the Red Maasai ($P<0.05$).

When any value of ei is less than 5, the above formula should be replaced by:

$\Sigma(|oi - ei| - 0.5)^2/ei$

where $|oi - ei|$ is the positive value of the difference between oi and ei.

This is because the c2 test is only an approximation to the distribution of numbers in a contingency table when the numbers are small.

Planning experiments : Careful planning of any experiment is obviously important, but it is even more essential for certain types of livestock trials and for on-farmexperiments. Experiments involving calving, lambing or kiddingmay be particularly awkward since it may be difficult to control birth dates. This means that different animals may enter the experiment at different times. In addition, livestock trials are often fairly long term, when compared to crop trials, and need intensive data recording. Animals are also valuable. This makes detailed planning essential.

□□□

Chapter 11

Appendix

11.1. Atomic Weights (Based on the Atomic Mass $^{12}C = 12$)

Name	Symbol	Atomic Weight	Name	Symbol	Atomic Weight
Actinium	Ac	227.0278	Neodymium	Nd	144.24
Aluminum	Al	26.9815	Neon	Ne	20.183
Americium	Am	(243)	Neptunium	Np	237.0482
Antimony	Sb	121.75	Nickel	Ni	58.69
Argon	Ar	39.948	Niobium	Nb	92.906
Arsenic	As	74.9216	Nitrogen	N	14.0067
Astatine	At	(210)	Nobelium	No	(259)
Barium	Ba	137.27	Osmium	Os	190.2
Berkelium	Bk	(247)	Oxygen	0	15.9994
Beryllium	Be	9.0122	Palladium	Pd	106.42
Bismuth	Bi	208.980	Phosphorus	P	30.9738
Boron	B	10.811	Platinum	Pt	195.08
Bromine	Br	79.904	Plutonium	Pu	(244)
Cadmium	Cd	112.40	Polonium	Po	(209)
Calcium	Ca	40.08	Potassium	K	39.098
Californium	Cf	(251)	Praseodymium	Pr	140.907

Carbon	C	12.01115	Promethium	Pm	(145)
Cerium	Ce	140.12	Protactinium	Pa	231.0359
Cesium	Cs	132.905	Radium	Ra	226.0254
Chlorine	CI	35.453	Radon	Rn	(222)
Chromium	CR	51.996	Rhenium	Re	186.2
Cobalt	Co	58.9332	Rhodium	Rh	102.905
Copper	Cu	63.546	Rubidium	Rb	85.47
Curium	Cm	(247)	Ruthenium	Ru	101.07
Dysprosium	Dy	162.50	Samarium	Sm	150.36
Einsteinium	Es	(252)	Scandium	Sc	44.956
Erbium	Er	167.26	Selenium	Se	78.96
Europoium	Eu	151.965	Silicon	Si	28.086
Fermium	Fm	(257)	Silver	Ag	107.870
Fluorine	F	18.9984	Sodium	Na	22.9898
Francium	Fr	(223)	Strontium	Sr	87.62
Gadolinium	Gd	157.25	Sulfur	S	32.066
Gallium	Ga	69.72	Tantalum'	Ta	180.948
Germanium	Ge	72.61	Technetium	Tc	(98)
Gold	Au	196.967	Tellurium	Te	127.60
Hafnium	Hf	178.49	Terblum	Tb	158.924
Helium	He	4.0026	Thallium	Ti	204.38
Holmium	Ho	164.930	Thorium	Th	232.038
Hydrogen	H	1.00794	Thulium	Tm	168.934
Indium	In	114.82	Tin	Sn	118.710
Iodine	I	126.9044	Titanium	Ti	47.88
Iridium	Ir	192.22	Tungsten	W	183.85
Iron	Fe	55.847	Unnilhexium	Unh	263.1182
Krypton	Kr	83.80	Unnilpentium	Unp	262.1138
Lanthanum	La	138.91	Unnilquadium	Unq	261.1087
Lawrencium	Lr	(260)	Unnilseptium	Uns	262.1229
Lead	Pb	207.2	Uranium	U	238.03
Lithium	Li	6.941	Vanadium	V	50.942
Lutetium	Lu	174.97	Xenon	Xe	131.29
Magnesium	Mg	24.3050	Ytterbium	Yb	173.04
Manganese	Mn	54.9380	Yttrium	Y	88.905
Mendelevium	Md	(258)	Zinc	Zn	65.39
Mercury	Hg	200.59	Zirconium.	Zr	91.22
Molybdenum	Mo	95.94			

A value in parentheses represents the atomic mass number of the radioisotope of longest known half-life.

11.2. Molecular and equivalent weights of important chemicals

Chemical	Molecular weight	Equivalents weight
Sodium carbonate (Na2CO3)	105.9890	52.9945
Oxalic acid (COOH) 2. 2H2O	126.0666	63.0333
Hydrochloric acid (HCn	36.4610	36.4610
Sulphuric acid (H2SO4)	98.0776	49.0388
Nitric acid (HN03)	63.0129	63.0129
Sodium hydroxide (NaOH)	39.9972	39.9972
Potassium permanganate (KMn04)	158.0376	31.6075

11.3. Abbreviations of units in commonn use and their equivalents

Units	Abbreviations	Euuivalentfs)
(A) Weights		
Microgram	μg	0.001 mg (or) 0.000001 g
Milligram	mg	1000 μg (or) 10^{-6}g
Gram	g	1000000 μg (or) 1000 mg
Kilogram	kg	1000000 mg (or) 1000 g (or) 2.2046lb
Ounce	O.A.	28.35 g
Pound	(Jb.J	453.6 g (or) 0.4536 kg (or) 16 Oz
Metric tonne	MT	1000 kg
Quintal	q	100 kg
(B) Measures		
Microlitre	III	10^{-6}L (or) 10^{-3} ml
Millilitre	ml	10^{-3} L(or) 1000 μl
Centilitre	cl	10^{-2}L
Decilitre	dl	10^{-1}L
Litre	L(l)	1000000 μl (or) 1000 ml (or) 100 cl (or) 10 dl
Ounce	Oz	29.573 ml
Gallon	gal	3.785 1 (L)
(C) Concentration		
Milli equivalent	mEq	-
Equivalent	Eo	-

Micromole	Il mol	10^{-6}mol
Millimole	mmol	10^{-3} mol
Parts per million	ppm	µg/g (or) mg/kg
Parts per billion	ppb	µg/kg
(D) Length		
Inch	In	2.5399 cm
Foot	ft	0.3048 m
Yard	yd	0.914 m
Micrometer	µm	10^{-6} m
Millimeter (cubic centimeter)	mm(cc)	10^{-3}m
Centimeter	cm	10^{-2}m
Decimeter	dm	10^{-1} m
Meter	m	1000 mm (or) 100 cm
Kilometer	km	1000 m
Angstroms	Ao	10^{-10} m (or) lOOµµ
Units	Ab brevia tions	Equivalenta(s)
(E) Energy		
Units	Abbreviations	Equlvalent(s)
Calorie	c (cal)	4.184 J
Kilocalorie	k cal	1000 cal (or) 4.184 kJ
Mega calorie	M cal	1000 kcal (or) 1000000 cal
Joule	J	0.239 cal
Kilo Joule	kJ	0.239 kcal (or) 239 cal
Erg	erg	10^{7} J
Horse power	hp	746 watts
Watt	W(~)	1 J/sec
Celcius	C	(F -32) × 5/9
Fahrenheit	F	(C × 9/5) + 32

11.4. Normal Values for blood constituents and enzymes

Particulars	Cattle	Sheep	Goat	Pig	Dog	Cat	Elephant	Horse	Fowl	Rabbit	Man
Ca(mg/dl)	9.4-12.2	11.4	10.7	11-11.4	9.8-11.6	8.22	6.90	12-13.4	9-12	5.8-14.4	9-11
P(mg/dl)	4-9	5.1-9	3-11	4-11	2.2-4.0	5.3-7.6	3.2	2.4-4.2	4-8	4.16	2-4
Mg(mg/dJ)	2-5	2-5	2-3	2-3	2-3	2-3	2.2	2.5	5.9	–	2-4
Total Cholesterol (mg/dl)	50-230	64-120	55-200	152-154	125-250	95-135	94-117	96.8	58-94	5-71	150-250
Blood Glucose (mg/dl)	35-55	35-54	45-60	65-95	70-100	60-100	42-78	60-100	125-275	50-90	70-100
NPN(mg/dl)	20-40	20-38	30-44	20-45	17-38	–	45.5	20-40	20-35	–	25-40
Creatinine (mg/dl)	1.0-2.0	1.2-~.9	0.9-1.8	1.0-2.7	1.0-1. 7	1-2	1.6	1.2-1.9	1-2	0.8-2.6	1-2
Bilirubin(mg/dl)	0.01-0.5	0.1-0.5	0-0.1	0-1.0	0.1-0.5	0.15-0.5	–	1-2	–	0-1.0	–
Urea N (g/dl)	6-27	8-20	13-28	8-24	10-20	20-30	16.2	10-20	0.4-1.0	2.4	9-15
Total Protein (g/dl)	7.56	5.38	6.67	6.40	6.1-7.8	6.4-8.0	10.7	6.7	4.7-6.1	6.5	7.5
Albuminll(/dl)	3.60	3.07	3.96	2.08	3.57	4.01	–	3.40	–	2.70	4.10
Globulin(g/dl)	3.96	2.31	2.71	3.32	2.63	3.57	–	3.30	–	3.80	3.40
Lactic acid (meq/l)	5-20	9-12	–	–	8-20	–	–	10-16	. 20-98	–	–
GOT/AST(U/L)	78-132	61.4	167-513	32-84	23-66	26-43	–	226-366	175	47	–
GPT/ALT(U/L)	14-38	–	24-83	31-58	21-102	6-83	–	3-23	–	79	–
GDH (U/L)	31	20	–	0	1-9	–	–	0-11.8	–	16	–
SDH (U/L)	4.3-15.3	5.8-27.9	14.0-23.6	1-5.8	2.9-8.2	3.9-7.7	–	1.9-5.8	0.5-2.0	–	–
LDH (U/L)	692-1445	238-440	123-392	380-634	–	–	–	162-412	636	64-124	–
IDH(U/L	9.4-21.9	0.4-8.0	–	–	0.4-7.3	2.0-11.7	–	4.8-18	–	137	–
CK (U/L)	–	0-2.9	0-2.5	–	–	6.2-19.5	–	0-3.6	–	125	–
ACTH (U/L RBC)	1270- 2340	640	270	–	270	540	–	450-790	–	–	–
AP (U/L)	0-488	68-387	93-387	118-395	20-156	25-93	–	143-395	482	120	–

About the Authors

Dr. A. Sahoo, Principal Scientist and Head, Animal Nutrition Division, Central Sheep and Wool Research Institute, Avikanagar, completed his B.V.Sc. & A.H. (1987-88) from Orissa University of Agriculture and Technology, Bhubaneswar and M.V.Sc. (1990) & Ph.D. (Animal Nutrition) (1994) from Indian Veterinary Research Institute, Izatnagar, U.P. He joined Agriculture Research Service (ARS) in 1993. He has more than 22 year of research experience in the field of 'Animal Nutrition' with specialization in rumen metabolism and manipulation, feed resource evaluation, assessment of nutritional adequacy, and clinical and therapeutic nutrition. He has published 80 research papers, 16 books/compendium/manuals, five book chapters, 61 general articles, 14 popular articles and 170 contributory papers. He has guided several post-graduate students. He is a member of several important Professional and Academic Bodies and Research Advisory committee.

Dr. S.K. Sankhyan did his B.Sc. (Ag.) from University of Rajasthan in 1986, M.Sc. and Ph.D in Animal Nutrition from Himachal Pradesh Krishi Vishva Vidyalaya, Palampur in 1988 and 1991 respectively. He was awarded University merit fellowship during M.Sc. and Ph.D. program. He joined Agricultural Research Services as Scientist in 1991 at Central Sheep and Wool Research Institute, Avikanagar and presently working as Principal Scientist Animal Nutrition and I/C ARIS Cell. He has 20 year of research experience and pursued the research in the area of nutrition of sheep and goats on pasture and mineral nutrition of field animals for the last fifteen years. He is presently conducting research in AICRP on "Improvement of feed resources and its utilization in raising animal, production". He has published 4 Books, 64 research papers in National and International Journals, 12 leads/invited papers and presented 60 abstract papers in National Seminars. He is life member of five National Societies of Animal Science.

Dr. S.A. Karim, Director, Central Sheep and Wool Research Institute, Avikanagar, completed his B.V.Sc. & A.H. and M.V.Sc. (Animal Nutrition) from Orissa Veterinary College Bhubaneswar in the year of 1974 and 1976, respectively and was awarded University Gold Medal for securing first position in M.V.Sc. (Animal Nutrition). He has completed his Ph.D. degree in Animal Nutrition from National Dairy Research Institute, Karnal in the year of 1991. He joined Agriculture Research Service (ARS) in 1978.

He has more than 32 year of research experience in the filed of ruminant nutrition with specialization in nutrition environment interaction, growth maximization of lambs, range managed small ruminants' nutrition and carcass evaluation. He has published 4 books, 10 book chapters,160 research papers, 50 invited papers, 20 popular articles and 210 contributory papers. He has guided several Ph.D. students. He is a member of several important academic bodies and as Board of Studies, Academic council, Research Advisory committee.